TEN PULSES OF EVOLUTION & THE SURPRISING NATURE OF EVOLUTIONARY TIME

First edition. May 4, 2020.

Copyright © 2020 Michael A. Susko.

ISBN: 978-1393464624

Written by Michael A. Susko.

I0486603

Table of Contents

To the rising of consciousness in all living beings..

And thanks to NovOntos for the use of his graphic, *Lamella.*

TEN PULSES OF EVOLUTION

& THE SURPRISING NATURE OF EVOLUTIONARY TIME

Michael A. Susko

AllrOneofUs Publishing

AllrOneofUs Publishing
Baltimore, Md & Huntsville, Al

FOREWARD

Michael Susko has written a brief but compact and tightly argued work on evolution that could be of service to professional biologists and an eye-opener for the rest of us. He shows that the four billion years of biological evolution can be organized in a logarithmic time line. Starting with the beginning of life four billion years ago, the timeline is halved repeatedly as two billion, one billion, five hundred million, and so on up to the first manifestations of humans 6.2 million years ago.

Change is accelerating and occurs in the context of regular logarithms. These ten periods, the "ten pulses" of the book title—are called "Major Nodes." The period before each *major node* is a *precursor period* in which the fossil-finds show increasing approximation toward the major change. The time of major change is followed by an *explicit period* in which the record confirms the change of *the major node*. Most significantly, at each of these nodes, there is a noticeable development in terms of increased complexity, mobility, sociality, and consciousness. This framework provides for biological students and scholars, a time-map on which any previous or future discovery can be located.

Throughout the book, Susko provides evidence from a very large number of scholars in archeological fossil-finds, molecular estimates, and chemical evidence. The work is enhanced by a glossary so that readers who are not professional biologists, but who have an interest in the process of evolution can gain an understanding. He does not express his personal beliefs except at the end in a brief conclusion. There he states that perhaps a form of "cosmic consciousness" has had a hand in all of this. His conclusion is reasonable, but the evidence stands independently from the conclusion. His final two sentences sum up a possible take-away from this study. After pointing out that world religions seem to be aware of the journey in time, he concludes: "But here we have brought the fruits of science, and the evolutionary witness for our past, to show that life has been continually gifted with greater freedom to move, greater awareness, and a greater capacity to love one another. We are invited to join in this journey and help build the universe."

Richard P. Mullin 10/28/2021
PhD, Duquesne University

PREFACE

This project has been several years in the making, as its scope is wide and the evidence detailed. From the start of life to the advent of humans, a novel use of logarithms will be employed to create an evolutionary scale. Logarithms, as a concept in math, are little taught to our young and one that we do not usually consider as adults. Yet the phenomenon of logarithmic growth is routine in nature and can be used to open a window to better understand the world. It is pervasive and not hard to understand in its essence. It means that numbers or values can change in exponential fashion, with radical and multiplicative increases, versus linear increases of just adding one by one. Thus, a tree has a single trunk, many limbs, then dozens of twigs, which in turn bear hundreds of leaves. It's similar to our arms, which have a single bone from the shoulder, doubles below the elbow, and expands into multiple bones in our wrist and fingers. Or, when a hurricane jumps from category three to four, it's not an incremental change, but packs considerably more punch than the single digit of increase would indicate. It is also like a car going from 30 miles to 35 miles per hour, doubling the impact of a crash.... We could go on.

As stated earlier, this work will consider the application of logarithms to measure evolutionary change over time. Our findings will be twofold: such change is pulsed, exhibiting radical nonlinear leaps, and that these pulses are logarithmically placed. In a way, this is a simple project, in which we keep *halving* units from the start of life to the advent of humanity, and look for nodes of nonlinear leaps. Remarkably, there are. Importantly too, the evidence around these pulses also follows a logarithmic pattern, in which there is a longer period of uncertain and suggestive evidence prior to the actual node of change, then a shorter period for which the evidence becomes confirming and leaves no doubt.

There was no initial *a priori* theory for which evidence was forced to fit into a strait jacket. Rather, the evidence suggested the pattern first. But once the general theory is found to be working, we are directed to search for more findings which are supportive. The vagaries of the world, however, mean that there will never be a perfect fit, and the evidence seems to wobble back and forth at times. Even then, results are remarkably suggestive.

So I offer you to go on a time journey that is logarithmically paced. In terms of this work, it means that the pulses of evolution are quickening at regular log intervals. You can be the judge of whether the pattern holds true. Do not let the math deter you, for I will make it clear how the calculations were obtained. For some, it may be just trusting that the numbers are correct, as if this is a mathematical "just so" story. However, a calculator in which you can put in the base two and then add a given exponential will easily produce the numbers found in the scale used here. Also, a "log base 2 calculator," which produces quick and easy results, can readily be found and used online.

Perhaps you think that the age of discoveries is over, that one can't come up with something new that might radically change our understanding of time and evolution. But I am afraid that as this work suggests, increases in our consciousness and awareness are not yet over. I invite you to take this journey and participate in this discovery. In doing so, you may help to make for the next step of evolution that has yet to be written.

Michael Susko, July 2021
M.S., Loyola College

INTRODUCTION

This work offers a novel way to map evolutionary change from the start of life to the advent of humans. Rather than using the traditional, linear scale in which the increasing pace of events bunches up toward the end, we employ a logarithmic scale which expands our resolution as we come to the present. This scale will allow us to see clearly the patterning of events throughout the span of evolution.

The use of the concept of logarithm should not deter us. In this work, it means we will simply halve the time to mark major pulses of evolutionary change. Thus, if the start of life is approximately four billion years ago, we go next to two billion years ago with the first nucleated cells, then to one billion years ago with the advent of complex multicellularity, and so on until the 10th node when we come to the start of humanity, about 7.8 million years ago.

If this mapping holds true, it means that significant evolutionary change is accelerating as we come to the present. Remarkably, the pacing occurs with logarithmic regularity, in that we identify major pulses of evolutionary change at regular halving or log intervals.

Importantly too, and confirming for our hypothesis, the evidence for change presents itself in logarithmic fashion around these nodes. That is, each Major Node has a *precursor period* in which fossil finds are increasingly approximating a grade-level change, followed by an *explicit period* which actually confirms the change.

We place the *Major Node* itself in the logarithmic middle between the precursor and explicit phases, in which we envision grade-level change to have occurred. As to why this logarithmic march towards greater consciousness has occurred with mathematical regularity, we reserve for our conclusion. The main goal of this work will be to establish the successive pulses of evolution and present the evidence which confirms such patterning.

Before elaborating, it bears repeating that the Major Nodes of the log scale basically *halves* units over time. Thus, our path is from the origin of life at 4 billion years ago (or 4 bya), to the eukaryotic cell at 2 bya ago, to complex multicellularity at 1 bya, to the first vertebrates at 500 million years ago (or 500 mya), two layers of mammalian evolution at 250 and 125 mya, then proceeding through three layers of primate evolution at 66 mya, 33 mya and 16 mya, until the emergence of hominins or humans, about 7.8 mya. A total of ten Major Nodes encompasses this evolutionary span, in which each makes for a qualitative leap in consciousness, paired with a greater mobility and social engagement in the world.

It is important to delineate the advantages in using this model to orient ourselves in evolutionary time.

(1) A logarithmic scale provides an easy mnemonic device of halving, leaving us with only a few key dates to remember. It provides a quick grasp of where we are in the temporal scheme of things and how it relates to the whole. If we throw out a date, say 260 million years ago, we know that we are just before the 5th node of 250 mya, the time of the first mammals.

(2) The model offers an expanding visual scale, in which we can easily view all the information. The events do not all bunch up toward the end of a linear scale, leaving us with a chart with little utility. At the same time, a log scale provides precision. Thus, if we consider a specific fossil, dated at 230 mya, we can say it's 10% log distance after the Major Node for mammals.

(3) The model, with its precursor period, embraces uncertainty and allows that uncertainty to give us information. We are not yet sure that mammals have arrived at 300 mya during the precursor period, but that dispute

tells us something. The model also embraces certainty, providing a bandwidth in which we expect affirming discoveries. Thus, by 210 million years, or 25% log distance after the Major Node, we expect confirmation for mammals.

(4) Thus, this model is broadly predictive. The log chart provides an immediate frame of reference for existing fossil discoveries, and if there are gaps, it suggests what we should expect to see. The scale also directs us to pay attention to stretches of time that we might easily gloss over.

Summing up, a logarithmic frame of reference provides us with a powerful tool to advance our evolutionary understanding.

Precursor Events, Major Nodes, and Explicit Events

A critical feature of this work is that the detailed evidence for the pulses of evolution is organized logarithmically, with *precursor events* prior to the Major Node and *explicit events* afterwards. With this in mind, let us review how each node will be described.

First, we will start with the time of the Major Node and a brief description of the accompanying grade-level change. Next, *precursor* events will be described which increasingly approximate the grade-level change. Precursor events include indirect indicators such as molecular estimates, chemical evidence, and fossils with a partial suite of traits. Events during this time are often confused finds and have a higher degree of uncertainty.

After the Major Node, explicit findings bring us increasing certitude that a grade level change has indeed occurred. Usually, a strong sign of such change is found within 10% log distance of the Major Node, with definitive evidence usually occurring by 25% log distance after the Major Node.

In-between the precursor and explicit periods, in the logarithmic middle, we hypothesize the point when evolution tipped to a grade-level change. The exact time of change will remain largely invisible, given the chances of fossil preservation for any particular point in time, and with only a few of the new type being present at first.

To summarize, we propose that evolutionary events are logarithmically organized in time. That is, major change is organized around a logarithmic middle, between precursor and explicit periods on either side.

To see how our terminology might fit into contemporary paleontology, let us consider the concepts of *stem* and *crown* groups. At the "stem origin date"--what can be called the first ancestor of a group--very few, if any, of the identifying traits can be identified. Then after a series of often confusing stem forms, we come to more certainty, and finally to a member of the *crown* group, fossils close enough to be identified with current, living survivors. This first ancestor of the crown group is actually called the *last common ancestor* of all the survivors.[1]

We can now relate this to the Major Node with precursor and explicit events. The precursor period can be seen as containing the initial stem groups, occupying a confusing, fragmentary period of evolutionary change. Along this early path of the stem ancestor, we see many fossil offshoots which went extinct. Turning to the explicit side of the Major Node, we find a substantial number of traits for the crown group. Often, toward the end of the explicit period, we find the first identified member of the actual crown group. As for the Major Node itself, it can be seen as the logarithmic middle between the stem and crown groups, as a likely place that a critical tipping point was reached.

A Note On the Mathematics Used in the Log Scale

Let us take a step back and explain how math is used to come up with the numbers used in our scale. To review, a logarithm is a number, the amount of times you multiply a base to come up with a given number. If you want to come up with the number 16, for example, using a base of 2, you would have to multiply it four times. Thus, 2^4 = 16. To repeat, the base of two must be multiplied four times (the exponential or logarithm) to make for 16. In this work we will use the base of two, rather than the more common base ten.

We can now show how logarithms are used to measure the placement of major evolutionary nodes, with precursor and explicit minor nodes on either side to frame the evidence. First, we need to calculate distances between Major Nodes using powers of 2, as in 2^2 = 4, 2^3 = 8, 2^4 = 16.... For the much larger figure of 4 billion (for the start of life), we can use 2 to the 32nd power (or 2^{32}) which gives us 4.3 bya. As this over shoots the 4 billion marker, we will drop a tenth of a power to $2^{31.9}$, a calculation that puts us precisely enough at 4 billion or 4.0007 bya. For the next Major Node of change we will mark down the exponential by one, as with $2^{30.9}$, which puts us at 2 billion, and so on.

The reason for using this method is not to find a complicated way to halve numbers. Rather, in terms of the nuts and bolts of a log chart, we can now place events precisely on a log scale. For example, consider the precursor period before the hypothesized start date of life at 4 Bya. We can add 1 /10 log distance, or .1 to the exponential, giving us 2^{32}, and --using a calculator--find the numerical value of 4.3 bya. Near this point we find light carbon from Canada at 3.25 bya, which may have been produced by life.

If we go to the explicit side of the Major Node, we find early fossils from life at 3.77 bya. This is matched closely to 10% log distance *after* the Major Node, which puts this minor explicit node at 3.73 bya. To calculate this distance, we simply subtracted .1 from the exponential which took us down to $2^{31.8}$--one-tenth log distance later in time--and the result of 3.77 bya. To summarize:

$2^{31.8}$*10% explicit node (3.73 bya) First fossils of life

(270 million years difference)

$2^{31.9}$◙ Major Node (4 bya) Hypothesized origin of life
(300 million years difference)

2^{32} *10% precursor node (4.3 bya) Isotopic evidence

Note that equivalent log distances, on different sides of the Major Node, are not the same linear distance. That is, one-tenth log distance *prior* to the Major Node is a longer amount of linear time than one-tenth log distance *after* the Major Node. In this case, 1/10 log distance prior to the Major Node is 300 million years long (from 4.3 to 4 bya), while the 1/10 log distance after the Major node is only 270 million years long (from 4 to 3.73 bya), or 30 million years less. To sum up, equivalent log distances are shorter in linear time the closer we come to the present. This means our log scale can be viewed as expanding the resolution of linear time--with shorter linear distances becoming more spaced out, the closer we come to the present. It will serve us well to keep pace with evolutionary events that are accelerating in time.

Outline of the Ten Pulses

For the sake of clarity and to help set the stage, I present an outline of the ten pulses of evolution covered in this work, putting in bold the first node in a series, an evolutionary burst which sets the stage for further elaboration. The abbreviation "*bya*" means billions of years ago, and "*mya*" means millions of years ago. The chart reads from the bottom up.

7.8 mya 10. 1st Humans/Hominins
15.7 mya 9. Hominoids (apes)
31 mya 8. Anthropoids (monkeys)

63 mya 7. 1st Primates
125 mya 6. Placental Mammals/Eutherian

250 mya 5. Mammals

500 mya 4. 1st Vertebrates

1 bya 3. Complex Multicellularity

2 bya 2. Eukaryotic/symbiotic Cells

4 bya 1. Life's Origins

In the main, this work will present evidence supporting ten logarithmically paced pulses of evolution. Our expectation for the model is that it will provide a general good fit for the evidence, one which proves useful to view the sweep and detail of evolutionary history.

Patterning of the Major Nodes

Not all Major Nodes are equal, and there is tentative evidence for a larger pattern. Specifically, we find series of overlapping fours, in which greater change happens on the first, then every fourth node.

1. The first node is an evolutionary burst which stands out in a remarkable way, such as the start of life or the first human. The first of a series represents a stage that has great evolutionary potential, to birth a new series of developments in which sentience/consciousness and mobile engagement take a qualitative leap.

2. A qualitative elaboration occurs, often involving size increase, along with increases in consciousness, range of mobility and more social life.

3. Amid more dominant life forms, a novel grade-level change occurs which is more subtle and hidden in nature, yet sets the stage for the next evolutionary burst.

4. The 4th (or the 1st again). Another evolutionary burst, for which we return to the first description.

On the Meaning of Evolutionary Pulses

While our survey largely focuses on the concrete evidence of fossils and their placement on a logarithmic scale, these pulses represent the more invisible change of sentience and consciousness over time. This highlights that life has an inner dimension that we do not directly see. Life's activity is not based solely on its reactions to an exterior world. Following the biologist Zak, life possesses an "information potential" which puts together motor and mental dynamics to build up a "self-image." As we follow the evolutionary pulses that allow for the emergence of humanity, we witness life showing continued increases in capacity for such sentience/ consciousness. Zak describes this as *"reflections upon reflections"* making for an "unlimited capacity for complexity, exponential increases that we see through the course of time."[2]

Thus, with a logarithmic scale, we expect to see evolutionary events accelerating in time, showing radical expansions of consciousness, an unfolding of ever greater depth and complexity. In a paradoxical way, we could say that time itself is expanding. Put another way, compared to a day in the life of a single-celled organism, ours is richer and more varied. As for what this means for our future, if the past provides our best clue, is that consciousness will have yet more logarithmic unfolding.

1st NODE: LIFE'S ORIGIN
4 BYA

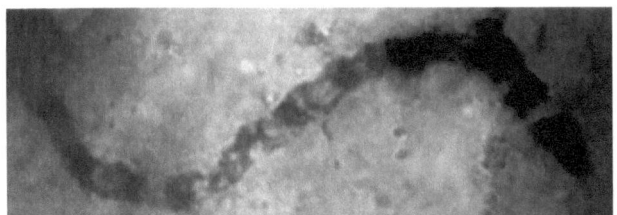

Procaryotic cells

The first log node is at 4 billion years ago, where we place the origin of life. No one would deny that the advent of life constitutes a great leap in the history of earth. Life has the ability to take in energy, tap the flow of entropy, then build up syntropy or order, and maintain a level of complexity that would not otherwise exist.[3] Its effect is so profound that it has created much of the atmosphere and altered the earth's landscape. An outline of events, based on a logarithmic scale, follows.

At the 1/3 precursor node (6 bya), the proto-sun developed, the incipient start of our solar system. After the 1/4 precursor node (4.77 bya), the solar system formed with the earth coalescing at 4.57 bya, a date inferred by the oldest meteorites. At 15% prior to our hypothesized start of life, the habitable boundary of life has been dated to 4.5 bya.[4] It's also the point when the continental crust developed, forming geological niches which furthered earth's chemical complexification. The earth's crust, forming and reforming repeatedly, was unstable at this early time due to high heat and continuous volcanic activity. Thus, it's generally theorized that early attempts of life would have been erased at this time.

At the 10% precursor node (4.3 bya), Zircons, a mineral artifact, are evidence for both water and supracrustal rocks.[5] A little later, the oldest dated rocks or crustal fragments have been dated to 4.28 bya. Thus, by this time, the earth had cooled sufficiently enough to have patches of crust, an atmosphere, and water. The rarity of the crust's survival and the earth's instability leaves us little hope for finding fossilized life, but chemical signatures offer clues. Zircon grains from Western Australia, dated 4.25 bya, possess high concentrations of light carbon, which may have been produced by life.[6]

Closer to the Major Node, to within 5% (4.15 bya), reports of light carbon in sedimentary rocks from Isua, Greenland indicate carbon fixation.[7] Dated 4.2 bya, organic matter with carbon isotopes consistent with life processes has been found in Western Australia.[8] These early signs are still best viewed as *potentially* showing the presence of life and insufficient to establish biogenicity with certitude.[9]

One more line of precursor evidence can be summoned. A recent genomic analysis, made possible by an ever widening availability of prokaryotic genome sequencing, provides evidence for life originating. In this study, life was found to have an origin in the Hadean age prior to 4.1 bya, and an early metabolism based on methanogenesis to have started between 4.1–3.8 bya.[10]

At the 4 billion node itself, a recent find from Canada shows graphite grains bearing light carbon, indicating the presence of life.[11] As expected, in the region of the 10% explicit node (3.73 bya), we have yet stronger signs, as chemical evidence becomes surer and more varied. In Greenland, dolomite formation formed by microbial mediation provides strong evidence for life at 3.7 bya.[12] Iron isotopes of pyrites also indicate life's presence in the Greenland formation.[13] Lastly, putative fossils, tubes and filaments, dated 3.77 bya, have been found near the earth's oldest hydrothermal vents.[14]

By the 1/4 explicit node (3.38 bya), we expect more sure confirmation. Fossilized life forms, dated ~3.5 bya, present communal, dome-like structures, strikingly similar to present-day stromatolites. The carbon ratios in the microfossil-bearing rocks are similar to later Carboniferous structures in which life forms have been clearly identified.[15] At 3.5 bya, methane of microbial origin, and filamentous single cells have been found in the same geological unit, evidencing microbial life with a different metabolism.[16] A little later, at 3.456 bya, a recent analysis from the Apex chert of Western Australia confirms the presence of life by shape and structure, as well as ion spectroscopy detecting carbon isotopes.[17] Such complexity at this early point is evidence that simpler forms occurred at an earlier time.

In the region of the 1/3 explicit node (3.16 bya), we find life inhabiting the deeper portions of the sea. Micrometer, mineralized tubes, and biogenic markers of isotopically light carbon have been dated 3.48–3.22 bya.[18]

Last, another whole field of evidence comes from South Africa, the Barberton Greenstone Belt, dated 3.55 to 3.20 Bya and spanning the one-fourth to one-third explicit nodes. This supergroup hosts "a large variety of convincingly biological macro- and microscopic, as well as geochemical evidence for early microbial life."[19]

In sum, the first Major Node placed at 4 billion years ago provides a reasonable estimate for the start of life, and for the framing for its evidence. During the precursor period, earth becomes ready for life and provides isotopic evidence for life's presence. The explicit period presents a variety of fossilized life forms. In general agreement with this, we note that the professional literature, as well as textbooks, 4 bya is increasingly cited as the date for the "last universal common ancestor of life."[20]

Log Scale of 1st Node, Life's Origin

*1/3 Explicit node (3.18 bya) Sulfide metabolism 3.235 Submarine life 3.48–3.22

*1/4 Explicit node (3.37 bya) Stromatolites 3.4 Apex Filaments 3.45 Methane microbes 3.5

Macro and microscopic life Barberstone Greenstone Belt 3.5-3.20

*1/10 node (3.73 bya) Isotopes indicate life 3.85 Putative fossils/hydrothermal vents 3.77

$2^{31.9}$ ◉ 1st Major Node (4 bya) START OF LIFE Molecular estimates 4.0 Light Carbon 4

*5% node (4.15) Light carbon, Greenland 4.15 Organic carbon, Australia 4.1

Genomic Evidence 4.1

*1/10 node (4.3 bya) Water 4.3–4.4 Oldest rocks 4.28 Light carbon from Australia 4.25

*1/4 Precursor node (4.77 bya) Solar system/Earth forms 4.57/4.54 Crust forms 4.5

*1/3 Precursor node (5.05 bya) Proto-sun forms 5.05

2nd NODE: EUKARYOTIC CELL 2 BYA

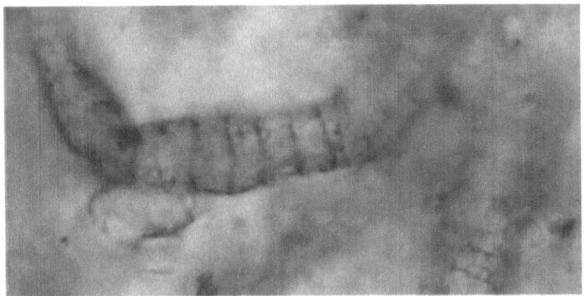

Eukaryotic filament

The second Major Node marks the arrival of eukaryotic cells, a life form which presents a leap in both size and complexity. The key to this evolutionary development is a novel symbiosis in which bacterial cells became incorporated within a larger host cell from the archaeal kingdom. It is well established that mitochondria were once independent microorganisms which became harnessed within a host cell. Critically, they provided an 18-fold increase in energy production. Major genetic exchange and alterations occurred because of the interaction between the symbiont and host. Additionally, sexuality is hypothesized to have evolved by this time, with profound consequences for life. Complex merging made for considerable variation, which served to vastly increase the repertoire of life.

With the Major Node falling at two billion years ago, let us now see how the evidence falls on a logarithmic scale. At the 1/4 precursor node (2.38 bya), earth was in the midst of the great oxidation event (2.4–2.35 bya), which has been characterized as "the most dramatic change in earth's history."[21] The earth's atmosphere was gradually enriched with oxygen, which became the energy source for the emergent symbiotic cell.

Typical of precursor periods, we find molecular estimates for the first eukaryotic cells with mitochondria, dated to 2.3 bya.[22] This date has recently been seconded by molecular estimates at 2.31 bya for sterol biosynthesis, a eukaryotic process.

Around this time, it's also believed that non-eukaryotic cyanobacteria, the most prolific producers of oxygen, evolved. Their presence, however, has only been definitely dated to 2.15 bya, approximately 15% log distance before the Major Node.[23]

Ten percent prior to the Major Node and up to the ¼ explicit node (2.1–1.7 bya), our resolution for preservation markedly improves.[24] It's during this time that microfossils lay claim to eukaryotic status.

Right at the Major Node (2 bya), large-size filaments called *Oscillatoriopsis* are possible eukaryotes, although size alone is not a sufficient criterion. The surest sign for eukaryotic status is the nucleus, which rarely fossilizes. Evidence for internal structures, however, is becoming available. In China, microfossils with eukaryotic structural traits of multilayered outer walls and visible spines have been dated to 2.15–1.95 bya. The Chinese team has ventured a genus association to *Dictyophera* and dubbed a new genus *Dongyesphaera*. From the Kola Peninsula in Russia, eukaryotes are reported based on the character of their cell walls and size. Dated 2.04 bya, they have been characterized as green algae.[25] At the same time a documented increase of oxygen in the marine environment would fit the scenario for eukaryotic emergence.[26]

Ten percent after the Major Node microorganism from Belcher Islands in Canada have been dated to 1.9 bya. The structures in these fossils compare to microorganisms deemed to be eukaryotes in assemblages half as old.[27] One planktonic microorganism, *Eosphaera* has been seen as related to the extant blue-green algae Eucapsi, as both produce a unique cubic colony.[28] These comparisons are seconded by the Canadian, Gunflint microbiota, dated 2.1–1.71 bya, in which *Eosphaera* and others are described as "probable remains of lower eukaryotes."[29]

Importantly, and also at the 10% explicit node, *Grypania*, with seaweed-like blades, has been dated to 1.85 bya. Its spirally coiled shape, a millimeter wide with pronounced regularity of appearance, argues for eukaryotic presence. However, no internal structures have been preserved to make for a definitive diagnosis.[30] A little later, at 1.8 bya, relatively large celled spheroidal phytoplankton may well be eukaryotic cells.[31] Another line of evidence, synthesizing genomic data and fossils, places eukaryotic emergence near the 1/10th explicit node. In this study, scientists concluded that the "integrated genomic and fossil evidence" leads to the estimate for the Last Eukaryotic Common Ancestor (LECA) to be at 1.84 bya.[32]

We should not be overly disappointed with the lack of certainty so far, given the immense time scale and our dependence on finding subtle microscopic traits. In addition, our model doesn't expect confirmation, but only strong signs within 10% log distance after the Major Node.

We must wait until the 1/4 explicit node (1.68 bya), for more certain fossil evidence. By 1.7 bya, several sites strongly indicate eukaryotic cells in terms of size, ornamented walls, and complex ultra-structures.[33] From South China, *S. macroreticulatum,* dated 1.7 bya, has been described as a eukaryote caught in the preparation stage for encystment.[34]

Halfway between the ¼ and 1/3 explicit node, at 1.63 bya, multicellular organisms from north China have been classified as eukaryotes, based on biosignatures, macro and microstructures.[35] From China and Australia, acritarchs, which resemble *Valeria lophostriata,* have been dated between 1.65–1.62 bya and are termed the "oldest unambiguous eukaryotic microfossils."[36] Frequently cited and dated at 1.63 bya, diversified unicells with cytological ultrastructures limited to eukaryotes have been paired with the eukaryotic species, *Tappania plana*.[37]

Moving closer to the 1/3 explicit node (1.59 bya), eukaryotic finds at a given site are not limited to a single species. Recent exploration of the Formations in China has identified 15 genera of potential eukaryotes, including filaments that look like red algae.[38] By 1.56 bya from North China, macroscopic multicellular eukaryotes, most likely red algae, have been identified, serving as a prelude to the next node.[39]

In sum, the second Major Node at 2 bya provides us with a good frame of reference for eukaryotic emergence. Precursor conditions are present 25% prior to the Major Node, with probable fossil eukaryotes occurring at the node itself. Verification follows with more refined structures and species identifications between the 1/4 and 1/3 explicit nodes. Last, we note the professional literature is increasingly offering 2 bya as a start date for eukaryotes, including a recent synthetic work that analyzed over 2,000 studies in a global "Time Tree of Life."[40]

Log Scale of 2nd Node Eukaryotic Cells

*1/3 Explicit node (1.59 bya) Undisputed eukaryotes *T. plana* 1.63

Acritarchs/ Accepted eukaryotes from Changcheng Formation 1.62

Valeria lophostriata (1.65) with macro and micro structures

*1/4 Explicit node (1.68 bya) Large acritarchs 1.7–1.8 *S. macroreticulatum* 1.7

Spheroidal phytoplankton 1.8

*10% Log distance (1.87 bya) *Grypania* 1.85 LECA ~1.84

$2^{30.9}$ ◙ 2nd **Major Node (2 bya) EUKARYOTIC CELLS** *Oscillatoriopsis* 2

Dongyesphaera /structural eukaryotic traits 2.15–1.95

Eukaryotic walls and size/Kola Peninsula 2.04

*10% Log distance (2.15 bya) Definite presence of Cyanobacteria 2.15

Preservation markedly improves 2.1–1.7

*1/4 Precursor node (2.38 bya) Biomarkers 2.31

Earliest molecular estimates for eukaryotic cells ~2.3

Great Oxygen event 2.45–2.32 Red beds appear 2.4

*1/3 Precursor node: (2.52 bya)

3rd NODE: COMPLEX MULTICELLULARITY 1 BYA

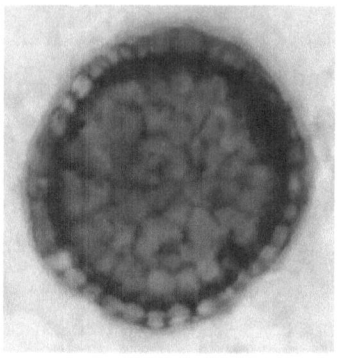

Multicellular Holozoan

The third Major Node at one billion years ago marks the continued evolution of macroscopic life and the emergence of complex multicellularity. The latter can be usefully defined as having clear cellular differentiation with more than one cell type, with multicellular interdependency.[41] This evolutionary step should not be underrated, and our logarithmic lens has not failed in directing our attention this way. After metabolism, multicellularity has structured this planet's biosphere, and represents a fundamental shift in individuality, out of which a host of new traits emerged.

Let us turn to the evidence. Around the one-billion year marker, the dominant life forms were algae, an in-between "kingdom" which is neither animal (although some ingest prey) nor plants with true roots, stems, or leaves. It is with this group that we find the first fossil record of complex multicellularity. If life before had been interactive mats, a new individuality emerges, in which many cells work together to make for an integrated organism.

As our focus is on the lineage leading to humans, however, we turn to the emergence of animals. The time of their first appearance still puzzles paleontologists. In terms of fossil evidence, several candidates for animals or metazoans have been proffered, but for each, a non-metazoan solution has been provided. We have said that the third node can present as more subtle and hidden, and this holds true for the origin of metazoans.

Let us start by scaling back our expectations for a full-blown multicellular animal at this node. Rather, we will begin by looking for an ur-metazoan among early unicells which manifest multicellular potential. Then hopefully, we will find evidence for complex multicellularity among organisms that are on the path to animals.

We know that all animals came from a unicellular organism, as evidenced by our own life cycle, which starts as a single cell. So we begin our search for the first animal by asking, from what group of protists did animals first evolve? Recently, this clade has been identified as *holozoans,* which also led to fungi, another multicellular kingdom, and three other types of unicellular protists. These other three protist groups are of interest because they may represent surviving links along the way to a multicellular metazoan. One of the three groups is the choanoflagellates, a unicelled organism which uses a flagellum to direct prey to its surrounding collar. Interestingly and of significance, choanoflagellate-like cells line the inner chamber of sponges, the oldest metazoan that is still extant.

We might ask: what is so different about holozoans that this clade harbored the potential to evolve the first multicellular animals? Recently, this group has been discovered to possess a genomic complexity qualitatively different from other protists. Thus, the first metazoan had only to retool genetic packets that the unicell already possessed.[42] These packets included novel genes for cell signaling, cell adhesion, and cell differentiation—critical tools needed to make complex multicellular organisms. Additionally, transmembrane receptors that trigger the animal's innate immunity response had evolved, a critical capability as pathogens would threaten novel multicellular life forms.[43] Last, holozoans possess the capability for coordinated contractions, dependent on molecular motors called actomyosin complexes, which are also found in animals.[44] In sum, the unicellular precursor to animals had already evolved considerable genetic complexity, which served as a pathway to complex multicellularity.[45]

Finally, we note that the unicellular precursor, rich in internal genetic complexity, markedly increased its complexity once the metazoan threshold was passed. Reconstruction of the ancestral metazoan genome reveals that an "unprecedented increase in the extent of genetic novelty" arose with the first metazoan, and is found in 25 groups of metazoan genes.[46]

Let us turn to the molecular and paleontological record to see how the Major Node at one billion years ago plays out. At the 1/3 precursor node (1.26 bya), early molecular estimates for metazoans can be found. For example, a recent molecular analysis placed the last common ancestors of Opisthokonta––the crown group of animals, fungi and their protist relatives––at 1.389–1.240 bya.[47] However, we note that many recent molecular estimates date much later, from 850 to 650 mya, between the 1/4 explicit node and mid-node.[48]

Turning to the fossil evidence, by the 1/4 precursor node at 1.2 bya, we find a rapid diversification of eukaryotic organisms. One hypothesis is that gendered sexuality made for this explosive radiation.[49] At 10% log distance before the Major Node (1.07 bya), the evolutionary burst had reached its endpoint, producing larger eukaryotes, greater in abundance, and with more morphological complexity.

At this time, the clearest example of a complex multicellular organism comes to light. Fossils attributed to *Bangiomorpha pubescens*, a red algae, have been dated to 1.047 bya, right at the Major Node.[50] These fossils exhibit true cellular differentiation and specialization, with a distinctive holdfast, cycles of cell division, spores, and sexually differentiated plants.[51] From another super-group, Chlorophyta or green algae, fossils that show a multicellular form with morphological differentiation, also present at 1 bya.[52] Lastly, multicellular fungi have been dated to 1.01–89 bya, in the region of the Major node. Fungal affinities include right-angled branching filaments, bulbous connections, and a terminal sphere.[53]

Thus, at the Major Node we have evidence for two major algae groups and fungal-like organisms attaining complex multicellularity. This confirms grade-level change at this node, albeit not with animals, our focus of attention.

We return to the protist, which became the first metazoan or animal. Here, the more fragile lipid membrane structures of animals present more challenge for preservation versus the tougher cell walls of plants or fungi.

Let us carefully frame this question. We ask: at what point did a proto-metazoan protist become multicellular with a complex life cycle, including development from a single cell? We know that undisputed fossils, indicating the presence of animals, have been dated securely 600–500 mya (the next Major Node). So, the question remains how far back we can take the stem-animal which exhibits multicellularity.

Returning to the molecular evidence, a recent comprehensive look at the super-kingdom from which animals arose, Opisthokonta, put the date at either 1.36 bya or 1 billion years ago. From that point, the timing of the first appearance of animals and their radiation have been calculated. As generally found in other studies, a *deep* pre-Ediacaran origin is supported, often between 1 billion and 720 million years ago.[54] This leaves us with a rather large bandwidth, and one that can only be resolved by paleontological evidence.

The general area around the one billion year marker, however, is not a vacuum, as we've already noted an extensive diversification of eukaryotic forms. These include multicellular types. In Arctic Canada, a newly reported assemblage dated 1.23–.9 bya, has 63 taxa and 25 eukaryotic types, along with three distinct forms of multicellularity.[55] There are also problematic microfossils, such as "complex multicellular vesicles," called acritarcha, which might be different developmental phases of a multicellular organism.[56] The question arises, do any multicellular entities at this time bear traits of the proto-metazoan, on its way to the animal type? Was there a "long stew" in this multicellular form becoming manifest, or a much later, short "popcorn" stage?[57]

Finding fossils of the ur-metazoan might be viewed as a near impossible task. Remarkably, a candidate has arisen. Recently, Strother has reported a complex multicellular holozoan from the Torridon Group in Scotland, which has been dated to 1 bya, right at the Major Node. Such rare life forms were revealed by newly developed techniques which do not extract the fossils from the rock matrix, but images them by electron microscopy to within a nanometer resolution.[58] With thick-walled exterior cells enclosing thinner walled interior cells, this fossil shows two distinct cell types, and stages for formation from a single cell to a multicellular structure.[59] Already the discoverers are comparing this reconstructed life cycle to one of the three extant holozoan unicellular groups, the Ichtyosporeans, in a free-living species, *Creolimax fragrantissima.*[60] In short, this multicellular freshwater protist has traits that are "consistent with a holozoan affinity."[61]

With at least two cell types, the organism rises to the bare minimum for complex multicellularity. Thus, this fossil find shows a potential pathway to the metazoan, in which gene-loaded unicells made for a multicellular organism one billion years ago. In one theory, *sequential* life phases became frozen *together* in time, making for a single complex, "novel spatiotemporal structure." With this vision of animal origins, cellular differentiation came first, then multicellularity arose.[62] In sum, evidence is present for complex multicellularity, a life cycle with differentiation, and the arrival of the ur-metazoan.[63] With multicellular development, a greater interiority of life arose, providing another logarithmic pulse in life's evolution.

This interiority also had a longer time to develop. Animals are noted for their extended life span, which includes jellyfish species which are immortal. In search of this "fountain of youth," researchers have identified a protein, p53, involved in extended lifespans and believed to be dated to 1 bya with the first holozoans.[64]

After the Major Node, we look for confirmation of metazoan origins. Recently, a strong signal has emerged with fossil finds from northwest Canada, microstructures which look exactly like earlier sponge body fossils found at the next node, the Cambrian era. Compared to usual animal finds, these are dated exceptionally early to 890 mya, about 15% after the Major Node.[65]

Early animal fossil finds have historically been problematic or suggestive. For example, a little later, at 900 mya, exceptionally large cells, known as megasphaeromorphs appear. These spherical, spheroidal or sausage-shaped structures have been interpreted as either colonial prokaryotes, eukaryotes, or metazoans.

By the 1/4 explicit node, metazoan affinities have been proposed with *Protoarenicola* and *Pararenicola*, although alternative explanations have been offered for each. With *P. Baiguashanesis,* annulated tubes with a bulbous terminal structure may be an alga, but the possibility remains that it is nearer to an animal. Its form has been compared to the Ediacaran, *Charaniodiscus,*[66] whose fronds and holdfast are like Cnidarian or coral-like animals.[67]

At the 1/3 explicit node (795 mya), more complex eukaryotes arose amid a Neoproterozoic radiation. Evidence for animals come in the form of chemical signatures, deposits of collagen, the glue that holds tissues together, dated >779 mya.[68] Lastly, and extending all the way out to the mid node, lipid products, dated 659–645 mya, provides concrete evidence for animals.[69]

In sum, the third Major Node presents evidence for complex multicellularity from multiple eukaryotic super groups. With metazoans, a suggestive holozoan with multicellular manifestations appears right at the Major Node. At 15% log distance later, we have reports of sponge microstructures. Possible fossil finds for metazoans are present by the ¼ explicit node, with confirming chemical signatures by the 1/3 explicit node. In terms of metazoan/animal evolution, we wish we had more definitive fossils for the ancestral animal. But the evidence is still unfolding, and our logarithmic lens has led us on a fruitful search for a multicellular animal at the one billion-year marker. We hypothesize that one day even more confirming evidence will arise. What we do know is that the stage has been set for an amazing radiation which occurred during the next node.

Log Scale of 3rd Node
Complex Multicellularity

*1/3 Explicit node (795 mya) Collagen deposits 779 Metazoan genes 800

*1/4 Explicit node (842 mya) *Protoarenicola* and *Pararenicola* Late molecular estimates for animals 850–650

*15% log distance (903 mya) Megasphareomorphs 900 Spores 900 Poriferan body microfossils 890 mya

*10% Log distance (934 mya) Molecular estimates for 1st metazoans 1.0–.72

$2^{29.9}$ ◎ 3rd Major Node (1 bya) MULTICELLULAR HOLOZOA 1.0

Multicellular Red Algae 1.0 Green Algae 1.0 Fungus 1.0

*10% Log distance (1.07bya) *Bangiomorpha* 1.047 Eukaryotic burst 1.1–.9

Molecular evidence for holozoan divergence

*1/4 Precursor node (1.2 bya) Rapid diversification of eukaryotic organisms 1.2–1

Arctic Canada diversity and multicellular types 1.23–.9

*1/3 Precursor node (1.26 bya) Molecular estimate for last common ancestor of Opisthokonta 1.389–1.240

4ᵗʰ NODE: VERTEBRATES
500 MYA

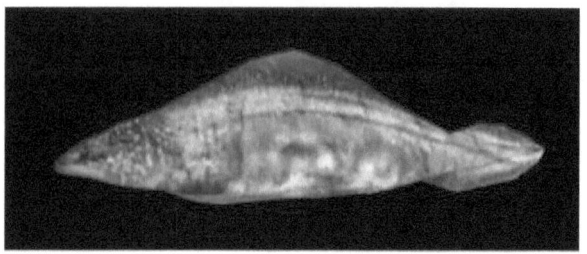

Haikouichthys

This node of evolution marks the fourth in a series, which promises to be revolutionary, and we are not disappointed with the advent of vertebrates. It presents a life form that is clearly animal, in terms of amplified movement, sensory organs, and a nervous system. More technically, a vertebrate can be defined as an animal having a brain enclosed by a skull, paired with a segmented spinal column.

This remarkable development represents a qualitative leap, highlighted by the innovation of the *head,* the vertebrate's most complex body part. The head has an array of sensory organs, the brain with critical neural processing, and a mouth/jaws—the site for initial food processing and sound production. Underlying the vertebrate innovation of the head is a novel cell type called neural crest cells, which arise from a fourth germ layer during development. These toti-potential cells have the ability to migrate during development to different parts of the body and make new structures.[70] Just as important as the head, we need to consider the dynamic body to which it is joined. Heightened movement results from a stiffened back structure made of bony vertebrae and joined with V-shaped muscles.

Let us turn to the fossil record and see how it aligns with a logarithmic scale around the fourth Major Node of life. With precursor events, we find sure confirmation for the presence of animals. Importantly, the expansion of obvious animal life was made possible by a significant rise in oxygen during the late Neoproterozoic era, 800–600 million years ago, between the prior half node and the 1/4 precursor node.

In the region of the 1/3 precursor node (631 mya), fossil evidence for animals appears. From South Australia, fossils dated 650–640 mya, have been painstakingly reconstructed from three-dimensional calcified remains, revealing ellipsoidal organisms with channels which appear to be from sponges.[71] Biomarkers from demosponges have been dated prior to 635 mya, seconding the appearance of sponge-like organisms at this early stage.[72]

At 630 mya, helically coiled spheroids have been interpreted as animal embryos.[73] Later, at 609 mya, approaching the ¼ precursor node, we find *Caveasphaera* exhibiting animal-like embryos which have an internal in-folding process similar to gastrulation.[74] Critically, this latter trait meant that an outer and inner germ layer became available for organ construction. It became the basis for the primitive intestine, which enabled *macrotrophy* or the ingestion of other multicellular organisms.[75] The capacity for specialized digestion can also be seen as expanding the organism's sense of interiority.

Close to the 1/4 precursor node (596 mya), China presents a "sponge grade body fossil," dated 600 mya, with tubes in pristine condition, measuring 1 mm by 1 mm.[76] Five percent log distance later at 575 mya, the first four waves of the Ediacaran Biota arrive, starting with the Avalon Wave, which included bilateral forms. *Dickinsonia tenuis* appears at 558 mya, about 15% log distance prior to the Major Node. This creature has been associated with lipid biomarkers, a sure indication of animal life, making it the first confirmed macroscopic animal.[77] It's quickly seconded by *Kimberella,* a mollusk-like, tripoblastic bilateralian, dated 555–558 mya.[78]

Regarding vertebrate origins, we look for animals with bilateral symmetry, a strongly supported clade. Approaching the 10% precursor node, at 550 mya, *Spriginna* possesses bilateral symmetry and clear head-tail differentiation[79]––although other traits indicate an arthropod/trilobite association.

Just after the 10% precursor node, which lands at 537 mya, early stem vertebrates have been found. *Pikaia,* dated 532 mya, is a tiny (4 cm) bilateral form, with a notochord, a stiffened cartilage tube-like structure that is fluid-filled.[80] This rod originates from the third, in-between germ layer, the mesoderm, which again indicates a greater interiority, and is paradoxically paired to greater movement in the world. The notochord is complimented by V-shaped muscles and another innovation, the tail, which makes for a powerful propulsion system.[81] Finally, *Pikaia* has a head with peculiar appendages and eyes. This head/brain area is joined by a body-length nerve tube, ectodermal tissue that went inward, by *invagination,* making for a central nervous system. This confirmed stem chordate, an immediate vertebrate precursor, remains enigmatic in terms of its placement within current vertebrates.[82] Such uncertainty and enigmatic finds are typical of precursor periods.

Closer to the Major Node (halfway between 10 to 5%) at 525 mya, we find fossils for a crown Chordata, *Haikouichthys.* A high degree of detail is revealed: lobe-like head, paired eyes, nasal capsules, and a series of ten plates that may be vertebrate elements.[83] This early stem vertebrate presents a puzzling mix of traits contrary to expectation.[84] Such varied traits of soft-bodied chordates reveal enough diagnostic details, however, that in composite, inform us of the likely traits of the ancestral vertebrate.

The Tommotian age, dated 529–521 mya, reports a type of vertebrate called conodonts,[85] with more definitive presence established by 500.5 mya, right at the Major Node. These creatures possess traits of the early vertebrates, including the first mineralized tissue and elements in the pharynx area made of dentine tooth. These elements, however, are not related to the teeth and jaws that emerged in the main line of vertebrates.[86]

Within 5% of the Major Node, *Metaspriggina*, dated 514 mya, features many vertebrate traits: prominent camera-like eyes, a series of W-shaped muscles, and gill features which prefigure jawed vertebrates.[87] Based on the lifestyles of extant chordates, with pharyngeal slits to feed upon rich bacterial mats on the sea bottom, the first chordate was most likely an active filter feeder.[88]

The explicit period should bring us confirmation for the full vertebrate type. Near the 10% explicit node at 467 mya, we find definite vertebrates. Dated 470 mya, the earliest complete fossil fishes, *Arandaspis,* look like a tadpole. Traits include a bony, shield-like head, two tiny eyes, and two pineal openings on the top of its head shield, probably sensitive to light and involved with circadian rhythms.[89]

A little later at 460 mya, *Galeaspids* possesses a mineralized cranial inner or endoskeleton, a sure vertebrate sign.[90] The intricate patterning under its large shields detail complex neural structures in its head, such as the pineal gland. Importantly, a reorganization of cranial sense organs has occurred, a prerequisite for the evolution of a jawed anatomy.[91]

Prior to the 1/4 explicit node, which lands at 421 mya, the closest jawless relatives to the jawed fishes emerged. *Osteostracans,* dated to the start of the Silurian (443 mya), provide a fully documented specimen by the mid-Aeronian stage at 440.8 mya.[92] These "plated-skin" fish feature the first armored dermal skeleton, with a head shielded by polygonal plates (tesserae) and teeth.[93] If injured, these plates exhibited a capacity for regeneration.[94] Trace lines portray a detailed internal picture within the head, revealing distinct brain parts, including a telencephalon or forebrain.[95] Sensory capacity was no doubt enhanced, as evidenced by lateral lines, which likely detected subtle changes in water pressure and weak electrical fields. The ear has evolved horizontal, semicircular canals, providing motion detection in a three-dimensional space. Last, a greater capacity for directed motion has emerged with paired fins, pectoral appendages related to our arms, and a more powerful elongated, dorsal tail. The rule is that heads and greater neural processing evolves in tandem with "tails," or a greater capacity for motion.

A little after 440 mya, Gnathostomes present the entire suite of traits associated with the vertebrates: jaws which evolved from a pair of throat slits, a phosphatic internal skeleton, vertebrae, and paired appendages.[96] With these traits, the full propulsive force of the vertebrate had arrived. Jaws, in particular, made for a momentous change, enabling enhanced respiration and the expansion of food sources, which led to diversification. Jaw structures are among the most evolutionary labile or changeable, and early jawed vertebrates experimented with a variety of forms.[97] One of the first gnathostomes, the placoderms, is confirmed by the mid-Telychian, ~436 mya.[98] Their armor was concentrated in head shields in the front and more lightly scaled in the back. These fish sported several pairs of fins and a long, flexible tail. Surprisingly, placoderms have also been found to birth live young.[99]

Evolution does not pause here. Approaching the 1/4 explicit node itself at 421 mya, *Osteichthyans* appeared, the group from which bony fish and tetrapods evolved. The oldest articulated skeleton of a bony fish, dated circa 423 mya, is the lobe-finned fish.[100] Already a precursor of tetrapods, their system of bones and muscles ties their fins to the rest of the skeleton.[101] As we pass the ¼ node, the first lungfish can be found, dated 417–412 mya. Its esophagus could gulp and process air, a pre-adaptation to land.

Finally, extending to the 1/3 explicit node (398 mya), vertebrates have emerged onto land, having taken on traits of tetrapods. The earliest tetrapod tracks are dated 397 mya, with molecular evidence pointing to a divergence between 41 and 97 mya.[102] The earliest, reconstructable tetrapod is dated 372 mya, halfway to the mid-node. Lastly, four genera of complete or near-complete fossils of tetrapods are dated between 36 and 59 mya, near the mid-node.[103]

In sum, the 500 million-year marker for the 4th node on our log scale provides a good frame of reference for vertebrate evolution. Early metazoans, bilateral forms, and stem vertebrates appear in the precursor period, while clear vertebrates appear after the Major Node. Already by the mid-node the evolutionary transition to land has occurred. The scope of movement, from the tiny chordate forms of a few millimeters to much larger land forms with the full suite of vertebrate traits, documents this immense jump. As for the future, we sense that evolution has crossed a threshold, where we can sense the human form beginning to take shape.

Log Scale of 4th NODE
Vertebrates

+Mid-node $2^{28.4}$ (354mya) Complete fossils of tetrapods 365–359

Earliest tetrapod *Parmastega aelidae* 372 mya

*1/3 Explicit node (398 mya) Earliest Tetrapod tracks 397 Molecular evidence 416–397

*1/4 Explicit node (421 mya) *Osteichthyans*/Lobe Fin fish 422 Lungfish 412–417

Jawed vertebrates ecologically diverse 419

*20% log distance (436 mya) Placoderms 436 *Jawed Gnathostomes*/Suite of traits 440

K. Delectabilis 440.8 Osteostracons 443

Ordovician-Silurian extinction event 443.8

*10% log distance (467mya) *Galeaspids*/mineral endoskeleton 460 *Arandapsis*/earliest fish 470

$2^{28.9}$ ◉ 4th MAJOR NODE (501 mya) VERTEBRATES

*5% log distance (519) *Metaspriggina* c. 514

Haikouichthys Agnates/jawless 525 Conodonts c 529–521

*10% log distance (537 mya) Chordates/Stem vertebrates 532 Pikaia 530

*15% log distance (556 mya) *Spriginna* 550 *Kimberella* 555–558

Dickinsonia 1st macroscopic animals 558 Trace animal fossils 565

*20% log distance (575) Ediacaran Biota 571–541

*1/4 Precursor node (596 mya) Fossilized sponge 600 Caveasphaera embryos 609

*1/3 Precursor node (631 mya) Embryos 630 Sponge biomarkers 635

5th NODE: MAMMALS
250 MYA

Morganucodon

The fifth Major Node at 250 million years features the first mammals. There's not much doubt that mammals represent a qualitative leap in consciousness paired with more dynamic movement and a greater social capacity. Brain size and specialized neural fields increased greatly, as a new layer of brain, the neocortex, evolved. Sensation notably enhanced, at first with smell and the somatosensory domains, as the first mammals were nocturnal. Then visual and aural domains greatly expanded, as daytime activity ensued. An important transition occurred around 220 mya when the three-ossicled, middle ear—unique and diagnostic for mammals—evolved from excess bones of the old tetrapod jaw.[104] The evolution of the ear is linked with that of a new jaw joint, used as the key identifying trait of fossil mammals.

Improvement in the jaws was a critical component in mammalian evolution, as they made for more efficient food processing, which enabled endothermy or "warm-bloodedness." Regulating the body's temperature internally with endothermy entailed a tenfold increase in energy cost, and allowed for greatly elevated levels of activity.[105]

Related to this, the mammalian four-chambered heart is inferred to have evolved at this time, an organ which separated oxygenated and unoxygenated blood and could distribute energy to maintain high levels of activity.[106] Last, another key mammalian trait, which doesn't readily fossilize was glandular skin from which mammary glands evolved.[107] The mammalian mother's ability to pass on a high energy source to its young led to a close bonding and foundation for a greatly expanded social capacity.

We now turn to the fossil evidence to see how it matches up to our logarithmic scale. Near the 1/3 precursor node (316 mya), the group from which mammals arose, the synapsids—so called because they have *one* temporal opening in the skull--have been dated to 320 mya.[108] This group of mammal-like relatives branched from tetrapods before reptiles had evolved, meaning our evolutionary path did not go through a "reptilian" phase. With synapsids, we have the first intimations for the mammalian line, with "pelycosaur groupings" that are a grade-level mix of related tetrapods.

Just prior to the 1/4 precursor node which lands at 298 mya, the diversity of amniotes or egg-laying animals had expanded enormously by 305 mya--a time which birthed the mammalian direction. Pelycosaurs diverged into three suborders. One was the carnivorous Sphenacodontidae, dated 304 Mya and which included the famous sail-fin *Dimetrodon*, often mistaken for a dinosaur. Importantly, this suborder has been identified as a sister group to later fossils, the therapsids, our stem mammalian ancestors.

A little after the 1/4 precursor node, the therapsids evolved with the appearance of *Tetraceratops insignis*, dated ~284 mya. Historically this find has laid claim to being a basal therapsid, but recent opinion has gone back and forth on the issue, with the latest saying it's only a sister group.[109] At 10% log distance before the Major Node (268 mya), we look for a stronger signal for the mammalian type. Here, we can definitely place the first therapsids, dated 275–265 mya.

Therapsids constituted a revolution in terrestrial life, with clear signs of higher activity levels. Limbs are no longer in a sprawling posture, but are more gracile, longer, and *under* the body.[110] With a simplification of their humoral leg form and a reorganization of their pectoral girdle and forelimb, therapsids gained a range of motion in their forelimbs that was critical to their diversification.[111] Also important, a hallmark trait of mammals is the regionalized spinal column, an S shape that enables a variety of specialized functions. Recently, it's been proposed that pectoral differentiation associated with forelimbs evolved first, then lower back or lumbar differentiation.[112]

With a more varied capacity for movement, we expect parallels in neural elaboration. Evidence for brain size has been difficult to ascertain, however, due to the lack of fully enclosed fossil craniums with therapsids. By way of comparison, we find suggestive evidence with non-mammalian therapsids, who have more cranial covering, and show a general increase of brain volume, as well as a great variety of "neurological diversity."[113] To cite an example of convergent evolution, a neocortex-like brain structure has been found in *Kawingasaurus,* dated 259–254 mya. This digging therapsid, not on the direct line to mammals, exhibits a surprising encephalization quotient, or brain to body ratio, two to three times larger than others at this time.[114]

Returning to therapsids, two strong indicators have recently emerged that they were endothermic or warm blooded.[115] The first is a study of bone histology, indicating a high resting metabolic rate of therapsid, at >260 mya.[116] Second, oxygen isotopes found in bone and teeth also show higher body temperatures for late Permian therapsids.[117] All this meant that a higher level of activity could be sustained. Another indicator for an active lifestyle is the therapsid nose, which provided a cooling mechanism so that heat could be dissipated.[118] Last, we note that therapsids have been found in the colder regions of Gondwana, adjacent to large ice sheets, leading us to believe that an endothermic capacity was present.

Along with movement and neural enhancements, we look for evidence of a greater social complexity. With therapsids, we find prevalent "cranial display structures," ornamental features on their head casings. This has been interpreted to be related to fighting and displays in matters of sexual selection. A richness in reproductive behavior infers that dynamics between the sexes were involved in the origin of mammals.[119]

To recap and add one more thing, at least four main traits define the mammalian type.

(1) A restructured jaw with a new joint and a single lower dentary bone, with the other bones having migrated to form the inner ear.

(2) Resultant structures of the "three-boned middle ear" enhanced hearing, whether to hear one's young, or catch prey such as buzzing insects.

(3) The neocortex, a new layer of brain evolved, making for another level of information integration.[120]

(4) Least we leave out teeth, mammals have differentiated types, which have precise occlusion—upper and lower teeth matching each other--so that food is efficiently chewed. A lot of energy needed to be processed by the jaw and its associated structures, to maintain the mammalian high

metabolic rate and level of activity, as well as the more metabolically expensive neural tissue.

Returning to our log chart, within 5% of the Major Node, the cynodonts, a therapsid sub-taxa or clade, have been dated to the Wuchiapingian age, 259–254 mya. Early cynodonts also had features which made for better food processing. Their teeth had three cusps and some degree of occlusion. A secondary palate, a roof over the inner mouth, enabled animals to breathe and eat at the same time. A second jaw muscle, the masseter, made for a muscular sling which allowed for complex jaw movements.[121] Turning to the spinal column, the loss of lumbar vertebrate meant that a diaphragm could evolve, which enabled a greater degree of respiration.[122]

With cynodonts, a pulse of encephalization to about .2 EQ occurred. The brain/head included small olfactory bulbs, a featureless forebrain, and middle ear components still attached to the jaw bone.[123] The cochlear part of the ear, which detects frequencies, is small and globular and not yet extended.[124] Compared to the wide array of therapsid clades, the cynodonts were only a minor group, but they survived to flourish and birthed the mammals after the Permo-Triassic extinction at 252 mya.[125] Remarkably, and which we have thus far failed to mention, therapsids were small, their heads on average only 2 1/3 inches long.[126]

With the arrival of the Major Node at 250 mya, we come to the cynodont named *Thrinaxodon*, about the size of a fox. *Thrinaxodon's* mammalian features include:

(1) A more upright posture, from sprawling to erect, with hind limbs pulled in closer to the body. A new hind limb retractor, the gluteal muscle.

(2) Three types of teeth, which are typical of mammals: incisors, canines, and molars, the latter having several well-developed cusps.

(3) Changes in jaw muscles, enhancing power and the ability to move the jaw, so that the creature is able to *chew*.[127]

(4) The secondary palate, the plate of bone which separates the mouth from the nasal area, is largely complete by now. As stated earlier, this made eating and breathing possible at the same time, an essential trait for an animal with a high metabolic rate.[128]

(5) Eardrums, which are connected to two small lower jaw bones, facilitated the hearing of airborne sounds. By this time the small three bones of the middle ear, the post-dentary bones, have reduced in size and begun to drift away from the jaw joint, enabling more sensitive hearing.[129]

(6) Fossils of curled up juveniles and juveniles with adults indicate parental care and homeothermy, or the maintenance of a stable body temperature. Fossil aggregations of different age groups are interpreted as mammals exhibiting a more "common social life" and parental care than with the first cynodonts.[130]

In sum, if *Thrinaxodon* moves, chews, hears, and cares for its young, like a mammal, it is nearly a mammal.

As we pass through the Major Node, we come to a strong indicator near the 10% explicit node (234 mya). *Adelobasileus,* dated ~ 230 mya (the Carnian age), is commonly held to be the first mammal. Its skull shows numerous mammalian traits, leading Benton to say it is "very probably a mammal."[131] But with only the upper skull preserved, we don't have yet the key identifying trait of the new jaw hinge. One mammalian trait we do possess is a fuller enclosed and strengthened housing for the cochlea, the inner ear complex. More insulated from the sounds of the body, hearing could be more successfully directed to the environment.[132]

We must wait to a point between the 1/4 and 1/3 explicit nodes (211 and 199 mya respectively), for the most basal mammaliaform, named *Morganucodon* and dated ~ 205 mya. This shrew-size animal with pointy teeth and scampering agility, likely ate insects and caterpillars.[133] Enhanced movement was due to all four of its feet being beneath the body, and vertebrae allowing flexion and extension of the spine during locomotion.[134]

Regarding neural evolution, a second pulse of encephalization is recorded to be .32 EQ, almost 50% more than earlier cynodonts, inferring a grade-level change. The greatest region of expansion was from the olfactory bulb and the related olfactory cortex.[135]

The importance of the "ortho-retronasal olfaction" complex, associated with smell, should not be underestimated. It came with the development of the secondary palate, which created a new passageway for air to go in and out of the nose. Air could be sniffed and subtle odors detected, including animals in their various hormonal states. It complimented the limited sensation of taste, so that air traveling from the mouth into the sensory area of the nose made possible the experience of flavor. Such humble beginnings with smell are hypothesized to have prompted the elaboration found in mammalian brain evolution.[136]

With regard to hearing, the inner ear or cochlea elongated from its original globular shape, picking up more frequencies, probably extending into the higher range to better hear buzzing insects and mammalian young.[137] *Morganucodon* also provides the first evidence of the mammalian developmental pattern that includes a juvenile stage before adulthood, suggesting a period of more intense parental care.[138]

A second main find, dated 203 mya and closer to the 1/3 explicit node, *Hadroconium* presents a more complete suite of mammalian traits. It finally possesses the new jaw joint, making for greater bite force and more efficient chewing. The old jaw bones have become fully re-purposed, incorporated into the middle ear, as the malleus and incus, resulting in a greatly enhanced auditory capacity.[139] For the first time, the middle ear bones have lost their attachment to the jaw. The cochlea also extended in length, indicating the detection of more frequencies.[140] Finally, *Hadroconium* has a wider and more expanded cranial vault, housing an enlarged brain for its size.[141] A third discrete pulse in encephalization brought the EQ to .5, which might well have marked the presence of the new brain layer found only in mammals, the neocortex.[142] Thus, on many scores: jaw joint, hearing structures, and neural changes we have definite indicators for the mammalian grade. It all adds up––the three pulses of encephalization, a greatly expanded movement capacity, enhanced smell and hearing, along with evidence for family life––to make for a qualitative leap in evolution at the fifth node.

In sum, the fifth Major Log Node at 250 mya frames mammalian evolution well. The earliest divergence of the mammalian precursors, the synapsids, occurred at 1/3 log distance prior to the Major Node. By the 1/3 explicit node, fossils present enough key traits to identify a mammal. In-between, a host of changes are building to culminate in the mammalian type. Was there an exact point that the mammalian threshold was passed? The Major Node served us well here, presenting strong evidence for endothermy, and a reasonable case for mammalian arrival. A recent major mammalogy textbook agrees, placing the "earliest identifiable mammals," and the attainment of the "mammalian grade" at ~ 250 Mya.[143]

Log Scale of 5th Node
MAMMALS

*1/3 Explicit node (199 mya) Sinoconodon 193 *Hadroconium* 195

*1/4 Explicit node (211 mya) *Morganucodon* 205/ (Rhaetian age 208–201)

(more definite mammal)

*10% Log distance (234 mya) *Adelobasileus* ~230 (Carnian Age 237–227) (1st mammal)

3 pulses of mammalian encephalization/olfactory expansion 225–195

Mammalian 3 ossicles inner ear 230

$2^{27.9}$ ◙ **5th Major Node (250 mya) MAMMALS** *Thrinaxodon* 250

Permo-Triassic mass extinction 252

*5% log distance (259 mya) Cynodonts c. 260 mya *Kawingasaurus* w/large EQ 254–259

Evidence for endothermy ◈ 260

*10% Log distance (268 mya) 1st Therapsids 270–265 Mammalian traits emerging

*1/4 Precursor node (298 mya) Pelycosaurian-grade synapsids 298–307/common by 299

Sphenacodontidae/reflected lamina 304 Mya

(Suborder of Pelycosaurs/sister group to Therapsids)

*1/3 Precursor node (316 mya) Synapsid Tracks 315

Synapsida diverge from Eureptilia 318–333 (~320)

6th NODE: PLACENTAL MAMMALS
125 MYA

Eomaia

The 6th Major Node of evolution lands at 125 mya, a time not given much prominence in our evolutionary history. Once we direct our logarithmic lens this way, however, we are able to discern grade-level change with the arrival of placental mammals.

Already, we have an expectation that the third node of a series will be more hidden. Finds for mammals at this time are typically rare. The placenta, a key trait, is a soft tissue organ with an ephemeral nature and highly unlikely to be preserved. Thus, we need to look for other traits that evolved with this group and see what evidence can be summoned. Although both therian (the pouched mammals) and the eutherians (typically termed placental mammals) have placentas, our discussion will focus on the latter group, as they were the lineage that led to the human line. They also have the most intensively developed and sophisticated placenta with a much longer phase of maternal retention.

Before proceeding to specific fossil finds, let us outline the general features which show that substantial change occurred around the 6th node. First, we consider movement and that a lifestyle dedicated to tree dwelling emerged. An arboreal mammal evolved a complex suite of traits, such as forearm pronation, which enhanced movement in trees and enabled a greater manipulation of things.[144] A second key area is sensation, in which significant gains in hearing occurred. This involved the cochlea becoming longer and more coiled in the inner ear, enabling eutherian/placental mammals to hear high frequencies.[145] More complex movement in a three-dimensional world, along with an expanded range of hearing would, in all likelihood, be accompanied by a pulse in neural evolution. Such organizational changes are not readily preserved in the fossil record. Nonetheless, we can reasonably hypothesize that a greater capacity to hear would lead to enlarged brain areas to process the enhanced sensations. Likewise, moving in the complex realm of fine branches with its attendant challenges would prompt motor areas in the brain to develop. This is verified by cladistic analysis showing that in contrast to marsupials, placentals evolved a novel motor cortex (M1) and premotor areas.[146]

Another neural innovation is specifically credited to eutherian mammals: the *corpus callosum*, which are fiber tracks that directly connect the brain's two fissured neo-cortexes. Earlier networks exist, but do not grant such direct connectivity. In terms of function, the corpus callosum is believed to be involved in integrating various sensory domains, as well as abetting faster movement.[147] It may also be the basis for more open-ended associations, which lays the foundations for symbolic consciousness.[148]

Let us now turn to specific fossil finds and see how they match up on a logarithmic scale. Setting the stage for eutherians is the evolution of tribosphenic teeth. Dated deep in the precursor period, between the mid-node and the 1/3 precursor node, we can place *Ambondro*, the oldest tribosphenic mammal of Bathonian age, ~167. Teeth are not to be underestimated in their evolutionary importance, as they open up new food sources which prompt diversification. With three-cusped molars, a more three-dimensional surface was precisely fitted or occluded with the teeth opposite. A crushing and shearing sideways motion[149] greatly enhanced the initial stages of digestion, expanding the surface area for the animal's bacterial biome to release nutrients.

By the 1/3 precursor node (158 mya), early claims for a eutherian appear at 160 mya. *Juramaia* has eutherian traits in its teeth, shoulder, and forelimbs. Phalanges/fingers have adaptations for scansorial/climbing, completing a portrait for an agile, lithe mammal up in trees.[150] In terms of body weight, *Juramaia* is quite small, 15 to 17 grams or about 6 ounces. The precise dating of these "floating specimens" is not assured, however.[151] Doubt has been raised, as there is a 35 million-year gap between *Juramaia* and similar specimens. Recent analysis of molecular data in tandem with fossils, provides circumstantial evidence that *Juramaia* could be dated to 126.4 mya, right at the Main Node. Finally, questions have been raised whether this find belongs to the eutherian/placental class at all. Such varied doubts are expected in the precursor period.

At the 1/3 precursor node, we can place a mammaliaform from the widespread docodont family, *Agilodocodon scansorius,* dated to the late Jurassic, ~160 mya.[152] With a head the size of a penny, its body shows clear arboreal features, such as a greater range of ankle mobility. The dentation of this tree dweller appears to be omnivorous, and it may have resourced gum from trees.[153] We emphasize that *A. scansorius* provides more secure evidence that the second mammalian pulse of evolution included an arboreal dimension. For all small mammals, climbing was important, whether on variegated surfaces on the ground, or if they ventured in trees. This is reinforced by yet another fossil find at this node, *Henkelotherium guimarotae,* dated to the Kimeridgian age, 157–152 mya. Features included an elongated tail, suggesting a "grundplan" able to access a wide array of three-dimensional surfaces.[154]

Near the 1/4 precursor node (149 mya), we can date *Dryolestes* at 152 mya. Here, as with other fossils, new non-invasive imaging tools have granted us access to subtle structures within the head. This creature's inner ear possessed a 270 degree or 3/4 turn of the cochlea, a precursor condition for the fully coiled cochlea.[155] This organ also displays sophisticated innervation, translating into higher resolution of hearing and a greater frequency range.[156]

At the 1/4 precursor node, dental characteristics support Eutherian lineage among two newly found mammals, *Duristotherium* and *Duristodon*.[157] Fossils of mammals from these early times are generally restricted to teeth and jaws and occasional skulls, but new finds allow us to see how movement capacity was expanding in the mammalian lineage. A little after the 1/4 precursor node, we can place *Akidolestes cifellii*, dated 145–140 mya. This species exemplified one of the varied types of mobility emerging at this time, possessing both ground and arboreal features.[158]

Right before the Major Node, *Vincelestes* has been dated to 130 mya. This eutherian also possessed a cochlea with a coil of 270 degrees, but was 1/3 longer than early mammals, the extra length adding yet higher frequencies.[159]

Right at the Major Node itself at 125 mya, we are fortunate that a major find has surfaced in China. *Eomaia* presents an exceptionally complete skeleton, with outer ear and hair markings or evidence of fur. The pointy teeth of this 6-inch long, 1.2–1.5 lb. mammal meant that it was most likely an insectivore. The discoverers classified *Eomaia* as eutherian, as it has the typical dental formula, with wrists and ankles possessing numerous traits consistent with the clade.[160] Although initially identified as scansorial—intermediate between ground and tree dwelling—*Eomaia* may have been fully arboreal. Finger proportions and curvature of grasping feet, for example, mirror that of extant arboreal mammals.[161] While some morphological questions have been raised,[162] *Eomaia* is generally believed to be a stem eutherian. However, this is still contested as some paleontologists classify them to be stem therians.[163] The epipubic bone is present, a marsupial trait which indicates that full placentation may not have been achieved.

Recently, *Ambolestes zhoui* presents another major find from the same formation in China, also dated to 126 mya. The nearly complete skeleton of this eutherian presents new anatomical features such as the hyoid apparatus, a structure which roots the tongue.[164]

Yet other fossils from this time show that mammals had expanded into varied ecological zones, including the aquatic *Yanocodon* and a glider, *Volaicotherium*. As mammals stepped away from their original nocturnal and burrowing habits, which emphasized smell and somatosensory zones, they came into open spaces which demanded more hearing and visual acuity. It bears repeating to say that higher level neural processing necessarily followed to process and integrate these senses, a capability which has been described as multi-modal.[165]

We turn to the explicit period in hopes of finding more clear confirmation for grade-level change. About 10% log distance after the Major Node during the Early Cretaceous ~120 mya,[166] *Prokennalestes* from Mongolia has been classified as a basal eutherian, a "stem placental mammal."[167] Enough teeth and upper and lower jaw parts have been found to evaluate morphological and size variability.[168] The cochlea is 4 mm in length, twice that of early mammals, and has now coiled a full 360 degrees.[169] This "full coiling," described as a "key evolutionary innovation," is inextricably linked to a neural structure, the cochlear ganglion.[170] At this point we can definitively place the evolution of high frequency hearing at >20 kHz. Ultrasonic hearing is unique among mammals, although it has been lost in humans.[171] Not only would greater hunting abilities result, but we surmise it enhanced communication in the mammalian family. Last, we note that brain size is similar to extant small placental insectivorans.[172]

Ten percent after the Major node, at the Albian age ~ 113 mya, *Sasayamamylos* presents the earliest known example of modern mammalian dentation.[173] This species comes from a group, Asioryctitheria, described as the earliest modern mammals.[174]

In the region of the 1/4 log explicit node (105 mya), unusually complete dental specimens have been found with *Montanalestes keeblerorum,* dated to the Early Cretaceous, Albian age 100–110 mya.[175] *Montanalestes*'s eutherian nature, initially deemed "likely," has been recently confirmed by phylogenetic analyses,[176] and is now described as "uncontested."[177] By this time, the full complement of mammalian hearing capacity has been attained, as a delicate middle ear and a coiled cochlea resembles modern species.

The 1/3 explicit node lands at 99 mya, and around this time we find fossils from Uzbekestan and the first eutherian-dominated faunas, dated ~ 95 mya.[178] A little further out, *Kulbeckia,* dated ~92 mya, is described as a placental mammal and basal member of *Zalambdalestidae*—a group of eutherians showing remarkable convergencies or homologies to extinct and extant placentals.[179]

If we extend all the way to the mid-node at 90 mya, eleven "stem placental" taxa, known from cranial and dental elements, show that eutherians had achieved significant diversification by this time.[180]

In summary, the 125 million-year marker proves to be a useful frame for early placental mammalian evolution. At the 1/3 and 1/4 precursor nodes we have the first claimed fossil eutherians, with traits found in dentation, hearing, and arboreal lifestyle. In the region of the Major Node, relatively complete eutherians are present. At the ¼ explicit node we have uncontested specimens. By the 1/3 explicit node and into the half node, a confirming diversity of stem placentals arises. Consistent with this, Kaas, relying largely on the emerging field of phylogenomics or genomic comparisons, has recently put the divergence of placental mammals at 150 mya, at the ¼ precursor node, with the divergence into different orders at 100 mya, the ¼ explicit node.[181] With eutherians, a second major pulse of mammalian evolution has occurred, one that sets the stage for a whole new radical step in evolution, the first primates.

Log Scale of 6th Node
PLACENTAL MAMMALS

+ Mid-node (88.6 mya) *Eleven types of Eutherians 90*
Ordinal origination 85–100 (Bininda-Emonds)
*1/3 Explicit node (99 mya) *Pranyctoides* c. 91.5 *Kulbeckia* 94.3–84.9

First Eutherian dominated fauna 95

*1/4 Explicit node (105 mya) *Montanalestes Keebleri* Early Cretaceous 110–100

*15% log distance (113) *Sasayamamylos*/modern mammalian teeth Albian age ~113

*10% log distance (117 mya) *Prokennalestes,* Early Cretaceous ~120

Corpus Callosum/new motor cortex hypothesized ~117.5

Origolestes lii 123 detached middle ear

$2^{26.9}$ ◉ 6th Major Node (125 mya) PLACENTAL MAMMALS

Tree-climbing *Eomaia* 125

Sinodelphys 125 Oldest metatherian/marsupial

Ambolestes zhoui 126 Eutherian Juramaia/Molecular analysis 126.4

Ambolestes 126 Vincelestes 130

*1/4 Precursor node (149 mya) *Akidolestes*/Tree & ground dwelling 140–145

Durlstotherium & Durlstodon 145

Dryolestes leiriensis 152 Tribosphenic molars c 145

*1/3 Precursor node (158 mya) *Juramaia/Eutherian?* 158 mya?

Maiopatagium "mother of wings" ~160 mya

Arboreal *Agilodocodon scansorius* Late Jurassic ~160 Paleo

Megaconus mammaliaformis Callovian age 166–163 halo of fur

Major adaptive radiation 174–152

Peramurids (163–145/Late Jurassic) origins of eutherians

Henkelotherium/270 degree coiled ear 157–152

Ambondro Oldest Tribosphenic mammal (Bathonian age) ~167
+ Mid-node (177 mya) Early molecular divergence 180
Marsupial lineage split 173

7th NODE: PRIMATES
62.6 MYA

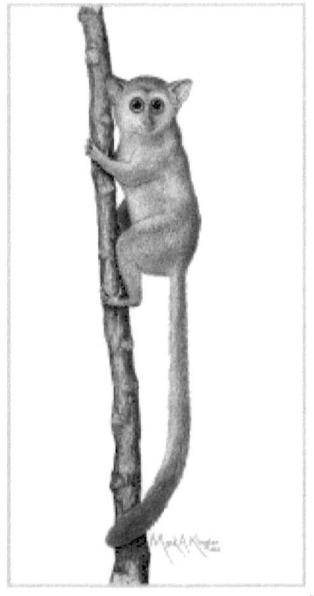

Teilhardina asiatica

The 7th Major Node features the appearance of primates, whose group goes on to make for two more pulses in evolution before humans. Primates certainly represent a qualitative leap in consciousness, indicated by a greatly expanded neural capacity, movement, and sociality. The technical definition of a primate, including fingers with nails and no claws and a post-orbital bar in the skull, does not fully capture the future this group will bring.

Although internal brain areas don't fossilize, we can extrapolate the differences that primates show compared to other mammals. As we might expect, primates have more unique visual areas (such as V3 and MT), as well as more modules and lamination in older areas carried over from mammals.[182] In tandem with this, early primates are inferred to have a greatly expanded posterior parietal cortex, which integrated visual and somatosensory domains to lead to action.[183]

Let us see how primate origins play out at the 7th logarithmic node, landing at 62.5 mya. This occurs right after a massive extinction, followed by a sudden radiation of modern-day placental mammalian orders. Of these, primates came from a placental group called Euarchonta, which includes colugos and tree shrews. These extant creatures exhibit arboreality and cranial features, strongly suggestive of the primate type.

We now turn to the evidence for primate origins. Immediately, we are confronted with the problem that early primate fossils have proved hard to find and are generally absent during the precursor period. At this time, we are largely left to estimates based on rates of change on the molecular level, which only gives us a general approximation. A recent review places the average age for molecular estimates to be at 82 mya, near the 1/3 precursor node (79 mya).[184] Another study used two strategies, and the genomic data created molecular estimates for 84.8 and 74.4 mya, near the 1/3 and 1/4 precursor nodes, respectively. In the 14 studies cited, the estimate for origins averaged to 71.7 mya, nearer the 1/4 precursor node.[185] Last, a robust molecular phylogeny recently gave ~71–63 mya as the range for the oldest ancestor of crown primates, between the 1/4 precursor node and the Major Node.[186] In sum, molecular estimates for primate divergence land in the precursor period, meeting the expectation with our logarithmic patterning.

We must wait until 66 mya, or 10% log distance before the Major Node, for fossil evidence. A group called Plesiadapiforms has its earliest date at 67.1 mya, with more definitive evidence by 66 mya. The first species distinction goes to *Purgatorius*, in which evidence from teeth place them as probable primates. Later finds of tarsals in the hindfoot area have verified their arboreal lifestyle and further primate affinities.[187] Generally, Plesiadapiforms are a non-leaping arboreal group, which radiated widely and successfully between 66 and 55 mya. They have a number of primate traits, mainly a similar dental array and climbing capabilities. Yet, Plesiadapiforms lack certain defining primate traits: a complete post orbital bar, nails on all or most fingers, and forward-facing eyes. The movement of these creatures is envisioned as slow and deliberate, relying more on touch and smell than sight.[188] Near the Major Node at 62.6 mya, the first skeleton for a plesiadapiform, from the Palaechthonid group has been dated to ~ 62 mya. This newly found evidence verifies an early tree-dwelling capability, interpreted as an "exclusively arboreal radiation of stem primates."[189]

Historically and among some experts currently, opinion remains divided on whether to call this group primates. Descriptions have varied from "primate relatives," "primate-like mammals" or "primatomorphs."[190] At a minimum, Plesiadapiforms are part of the Eurarchonta, the grouping from which primates emerged.[191] On the other side, cladistic analysis based on the fossil evidence has been claimed to "unambiguously" place Plesiadapiforms with true primates.[192] Of late, there has been a shift toward accepting the group as having obtained primate status, in which "the balance of evidence" supports the "stem-primate" hypothesis.[193] Again, we note that uncertainty is expected in the precursor region prior to the Major Node.

Ten percent log distance after the Major Node, the first claimed fossil evidence for "true primates," a primate with the full suite of primate traits, appears. Fragmentary remains of *Altialasius* from Mongolia, dated to ~58 mya, have been described as a very basal true primate.

A little before the 1/4 explicit node (52.7 mya) we find consensus for the emergence of euprimates or "true primates." A cranium found in China, named *Teilhardina asiatica,* and dated to 55.8 mya, marks the "debut of an undoubted primate." The skull of this small, mouse-sized primate possesses the strut enclosing the eye, orbital convergence of the eyes, and a relatively large brain compared to Plesiadapiforms. Weighing about a gram, an extremely small size for a primate, its convergent eyes may be evidence for daytime predation of insects.[194]

A little later at ~ 55 mya, we possess skeletal evidence for the earliest "dry-nose" (Haplorhine) type of primate, the ancestors of old world monkeys and great apes. These finds are also related to tarsiers, extant primates who are nocturnal and specialize in leaping. *Archicebus achilles* presents a surprising mix of tarsier-like traits and features which already foreshadow the next primate node, the anthropoids.[195]

At the 1/3 explicit node (49.7 mya), we find crown primates, species which can be identified with living members. At this point tarsiid-like fossils have been found, including *Xanthorhysis*,[196] and *Shoshonius* at 50 mya.[197]

In summary, the 63 million-year date for the seventh Major Node has provided a good fit for evidence of primate emergence. Molecular data places origins in the precursor period and within 10% of the Major Node we have a reasonable case for primate presence. By the 1/4 explicit node, undisputed fossil evidence for true primates appears. Finally, the 1/3 explicit node presents the first crown primates, which are like current species. With the advent of primates, the first of a set of Major Nodes, we usher in a new wave of evolutionary history.

Log Scale of 7th NODE Primates

*1/3 Explicit node (49.7 mya) Crown Primate/Tarsiids *Xanthorhysis* 50 *Shoshonius* 50

*1/4 Explicit node (52.7 mya) *Definite* primate/*Teilhardina asiatica* 55.8–50.3

Archicebus achilles /earliest "dry-nose" primate~5

*10% log distance (58.4) *Altiatlasius* ~58

$2^{25.9}$ ◉ 7th **MAJOR NODE (62.6 mya) PRIMATES** Palaechthonid 63.2–60.2

1st *Plesiadapiform* skeleton with arboreal traits

*10% Log distance (67 mya) *Purgatorius*/Oldest *Plesiadapiforms* 66

Major extinction event 66

*1/4 Precursor node (74.5 mya) Molecular est. average 71.7 Molecular est. 74.4

*1/3 Precursor node (79 mya) Outer range estimate 84.8

8th NODE: ANTHROPOIDS
31.25 MYA

Catopithecus

The first anthropoids cluster around the 31 million-year marker at the 8th Major Node of Evolution. These larger sized primates are characterized by technical-sounding traits: orbital closure of the eye, pneumatization of the inner ear, and a certain dental pattern. Of these, perhaps the most important diagnostic trait is the enclosure of the eye by bone, which helped stabilize the eye for acute vision.

In terms of a qualitative leap for this node, we can focus on three anthropoid characteristics:

1.) Acuity of vision, complimented by expansion of the optic areas of the brain and associated neural processing.

2.) A logarithmic expansion in body size, which meant that the neocortex likewise increased in size, number of neurons, and varied cortical fields.

3.) Sociality greatly intensified with troop dynamics, which in turn prompted more neural development. Indicators for increased sociality include visual acuity to recognize faces, sexual dimorphism, and observation of extant anthropoids.

In sum, major changes in vision/neural processing, greater body size, and sociality lead us to posit a qualitative leap with the anthropoid grade.

Let us turn to the fossil evidence and see how a logarithmic patterning holds. Surprisingly early and placed at the mid-node (45 mya), a major find presents from China a miniaturized primate weighing about 100 grams or 3.5 ounces. *Eosimias* is claimed to be a basal anthropoid, a designation which remains uncertain as there are no cranial remains. While some dental traits demonstrate anthropoid affinities, "several primitive characters" distinguish them from other fossil anthropoids.[198] Whether technically an anthropoid or not, the discoverer agrees that the grade-level change to anthropoids has not yet occurred. The find has been characterized as transitional, with traits between the earlier prosimian level and anthropoids.[199] All in all, the mixed nature of *Eosimias* matches our expectation for a mid-node, which lies in-between Major Nodes.

At the 1/3 precursor node (39.4 mya), a stem anthropoid *Amphipithecidea* from Southeast Asia has been dated at 40 mya.[200] While *Amphipithecidea* has been classified as a primitive anthropoid, fragmentary remains from teeth and jaws have been described as belonging to the earlier prosimian grade.[201]

The 1/4 precursor node (37.2 mya) presents the earliest "basal anthropoid" from Libya, dated 37 mya. *Biretia megalopsis* weighs 3/4 of a pound and has cranial remains bearing large eyes, suggesting they were nocturnal.[202] The primate shift to daytime vision had not yet occurred, but body size has nearly tripled when compared to prior primates.

After the 1/4 precursor node, *Catopithecus browni* from Egypt has been dated 34.8–33.9 mya.[203] This primate weighs a kilogram or 2.2

pounds, about three times the weight of the prior *Biretia*. The eye size of *C. browni* indicates a diurnal or daytime life style, and its body shows quadrupedal arboreal habits––traits nearing an anthropoid status. Its skeletal remains, however, are not typically anthropoid in all their features, presenting a mosaic.[204] Finally, orbital closure of the eye, a defining anthropoid trait, is lacking.

An important clue surfaces regarding the social life of these early stem anthropoids. Canine size differences or dimorphism between the sexes indicates social groupings, with male competition and polygynous mating systems.[205] In the sequence of teeth eruptions, canines came last, suggesting their importance in social display at a later point in this primate's life history.[206]

From the 10% precursor node through the explicit period, 34–21 mya, the fossil evidence has been characterized as poor, perhaps due to an extinction event around 32 mya. We are not totally lacking in fossil finds, however. At the 10% precursor node (33.6 mya), *Parapithecus grangeri* has been dated to 34 mya. Weighing in at 2–2.2 pounds, *Parapithecus* is believed to be a basal stem anthropoid. From its eye sockets and size differences in jaws and teeth, *Parapithecus grangeri* is considered to be diurnal and dimorphic.[207] A nearly complete cranium reveals that *P. Grangeri* had a relatively smaller brain for body weight compared to living primates. As for neural differentiation, a high resolution study found that retinal ganglion cell counts and olfactory bulb volume are intermediate between the living lemur-like strepsirrhines and monkey-like anthropoids.[208] These primates were sexually dimorphic in body and canine sizes, indicating they had a complex social life, in which groups of females associated with a single male.

Somewhere near the Major Node from 31 to 35 mya, the catarrhine group, which led to old world monkeys, great apes and humans, evolved trichromatic color vision. With three peaks in spectral frequencies interfacing, an array of colors could now be seen.[209] These stem anthropoids would not only be able to see the varied colors of leaves, flowers and fruits, but colorful displays from their peers, which served to heighten group cohesion.

By the 10% explicit node, we look for a stronger signal, and we find *Aegyptopithecus Zeuxis,* dated 30.2–29.5 mya. This primate, from the African Jebel Qatrani Formation, has been classed as a stem Catarrhine ("downward nosed") infraorder. This species has clear anthropoid features, including complete postorbital closure and dental formulae.[210]

A remarkable recent find of a nearly complete female skull of *Aegypopithecus* provides strong evidence for anthropoid status and a grade level change. Three lines of evidence can be summoned: (1) Expanded visual areas of the brain, the primary visual cortex, as measured by micro CT scanning. (2) Placement of eyes and relatively small orbits, indicating a daytime or a diurnal lifestyle, and (3) The presence of a marked sexual size dimorphism.[211]

Let us elaborate on the meaning of sexual dimorphism, in which males are markedly larger, and have other differences that distinguish them from females. Traditionally, it has been seen as indicating that males in a troop are competing intensely for females. But it could also be showing a reduction in intraspecific competition for resources such as food.[212] When compared to living primates classed as evolutionarily earlier, living anthropoids generally exhibit more secondary sex differences, whether in body size, canine size, color, or vocal calls.[213]

Other clues to enhanced sociality can be cited. The pattern of teeth eruption in *A. Zeuxis is* consistent with extant monkeys and a delayed development. Longer periods of immaturity allow for more social learning, as well as neural expansion.[214] Another indicator is the heightened range of frequencies heard, as cochlear length is now curled three times. This is seen as enhancing the ability to hear lower frequencies, which carry over longer distances, and is valuable for troop cohesion.[215]

Perhaps an underestimated change that occurs at this time is the great increase in overall size. The best current estimate for *Aegyptopithecus* is 6708 grams or 14.8 pounds,[216] six and a half times larger than the prior *Catopithecus*. Although *Aegyptopithecus's* frontal lobes are not as expanded as living anthropoids, the size increase of body meant that the brain's raw size has also greatly increased. In addition, *Aegyptopithecus* measures a higher Encephalization Quotient of .97, compared to the earlier *P. Grangeri's* of .77. Other brain changes likely occurred that are paleontologically invisible. Recently, it's been hypothesized that general neural size increase translates into an increased "corticalization of function,"[217]proportionally making for more neocortex, more neocortical areas, and more total neurons.[218]

At the 1/4 explicit node (26.3 mya), we find *Kamoyapithecus,* dated ~ 26 mya. This lightly built primate had thinly enameled teeth, a projecting jaw, and a small brain for its body weight. The identification of "dentally hominoid-like catarrhine" actually points toward the next node, although the level of preservation is insufficient to perform a cladistic analysis.[219] The find is what would be expected for a stem hominoid, but there remains an inability to find unambiguous, shared derived traits.[220] However these conjectures sort out, "dental apes" remain very similar to current Old World monkeys and are equivalent to monkeys from a functional and ecological view.[221]

As we approach the 1/3 explicit node (24.8 mya), evidence presents for the divergence between Old World monkeys and apes, again suggestive of the next node. Two paleontological finds, dated 25.2 mya, give early evidence of this future change. One is a tooth, identified as the oldest stem member of the extant African monkey clade. The other is identified as the oldest known fossil "ape," our next node, based on a partial mandible with a single tooth preserved.[222] Such slim finds do not yet confirm the next grade-level change, but at least we can posit sure anthropoid emergence.

In sum, sufficient evidence is present to establish anthropoids and a grade level change at the 7th Major Node. Qualitative change is illustrated in the domains of vision, hearing, body size, and sociality--all of which are matched by neuralization. If the earliest stages present an array of stem forms, the later stages are verging into the territory of "dental apes," which leads us to the third major pulse of primate evolution.

Log Scale of 8th Node
Anthropoids/Monkeys

*1/3 Explicit node (24.8 mya) *Rukwapithecus fleaglei* 25 Hominoid/ape divergence 25–23

*1/4 Explicit node (26.3 mya) *Kamoyapithecus/* "Dental ape" ~26 (27.5–24.3)

10% log distance (29.2 mya) *Saadanius hijazensis/* bony ear tube 29–28

Aegyptopithecus Zeuxis 30.2–29.5

$2^{24.9}$ ◉ 8th Major Node (31.3 mya) ANTHROPOIDS Trichromatic color vision 30–35

Propliopithecoidea 30 1st stem catarrhines/downward nose primates 32–29

10% log distance (33.6) *Aegyptopithecus Zeuxi* 33.9 /Large, sighted anthropoid

Parapithecus grangeri 34 Major extinction event 34

*1/4 Precursor node (37.2 mya) *Catopithecus* 35.7 Stem anthropoid

*1/3 Precursor node (39.4 mya) Molecular divergence of catarrhines 40–44

+ Mid-node (44.3 mya) *Eosimias* 45 / Basal anthropoid

9th NODE: HOMINOIDS
15.6 MYA

Dryopithecus

With hominoids at the 9th Major Node of evolution, we are on the cusp of human kind. Although sustained bipedality has not been achieved, large bodied primates are spending significant time on the ground. Posture is shifting to a more orthograde or upright position. Along with an increase in vertical climbing, there's more variety in mobility. Traits include well-developed grasping extremities, more mobile limbs, and greater elbow stability to maintain different postures. Greater time on land is also reflected in the loss of a tail, which had been useful in maintaining balance in trees.

With hominoids, the brain has increased in size, including critical front-white matter, which has proportionally increased. This "white matter" of the brain integrates the nervous system's feedback, furthering feeling, judgement, and self-consciousness.[223] The cerebella, responsible for movement, has undergone an allometric grade shift relative to the rest of the brain.[224] Allometry means that this particular area of the brain is growing in size more rapidly than the rest of the brain. Recent study has also confirmed the "exceptional expansion of the prefrontal cortex." It started with the great apes, dating ~19–15 mya, the area around the Major Node, before continuing on with humans.[225]

No doubt social life intensified, as we find with extant great apes. Modern day groups show fission-fusion social dynamics, in which individuals are able to leave and rejoin troop mates. At this point, a higher order of thinking may have evolved, mapping virtual reality, where the social brain goes from "what now" to "what if."[226] In sum, movement, neural capacity, and extrapolation from current great apes points toward a grade-level shift at this node.

Before exploring the fossil evidence, let us briefly consider a neglected great ape parallel to gain a better sense of this node. Using molecular evidence, the gorilla is hypothesized to have diverged at 15.1 mya, right at the Major Node.[227] This bypassed giant, who has weight and size comparable to humans, provides an unexpected wealth of parallels. Recent investigation of the scapula or shoulder blade suggests that humans are more related to the gorilla mode of *clambering* than modes linked to suspension from trees.[228] As the only great ape with a terrestrial life style, the gorilla nests on the ground, and moves on land with an unexpected variety of hand positions other than knuckle walking.[229] The gorilla is a generalist feeder, whose diet is mainly herbs unless fruits are available. This great ape has infant, juvenile, and subadult phases from 0 to 8 years. Later, at age ten, a sexual dimorphic stage emerges, with displays of red coloring on a crested head and silvering down the back.[230] With the gorilla and the other great apes, an array of research is confirming of grade level change, including the routine use of simple tools to extract food,[231] and the transmission of social traditions.[232]

Let us now turn to the fossil evidence in the precursor period. *Proconsolids,* stem apes who are absent a tail, already show a marked increase in brain size, while retaining anthropoid, quadrupedal locomotion. Proconsolid-like finds actually date back 23.5 mya, to the prior half node.[233] But the first generally accepted "archaic" *Proconsul* is dated 19.6 mya, at the 1/3 precursor node. The species, *Proconsul major* is the size of a female orangutan, has a relatively large brain, thin enameled teeth like African great apes, but otherwise had a dental formula like that of the earlier anthropoids.[234] A related genus, *Ekembo,* dated 19.5 mya, possessed a tail bone (meaning the tail was absent), more powerful grasping hands and feet, and a larger brain than expected. A little later at 19 mya, *M. clarki* possessed a short snout, large eye orbits, and a gibbon-like appearance. Thus, by the 1/3 precursor node, a variety of stem apes have appeared.

A little after the ¼ precursor node, *Proconsul helsoni* has been dated 17.9–17.5 mya. Possessing an ape-like face, this primate has a domed brain with a clear increase in encephalization quotient. Although *Proconsul helsoni's* movement remained quadrupedal, it had more upright posture and flexibility of movement, quite unlike any living ape today.[235]

Within 15% of the log node, *Afropithecus turkanesis,* dated 17.5 mya, still retains the older form of quadrupedal mobility. This is the first hominoid, however, to have thick-enameled teeth like humans. Also, *Afropithecus's* crown formation indicates three years of prolonged dental development and a delayed maturation.

At around the same time, a distinct genus, *Mortopithecus,* has been dated by associated fauna. While dentally mirroring the prior *Proconsul* group, the body's skeleton is more like that of the great apes. Its short waist is stiff for uprightness, and its scapulae can rotate the arm around the shoulder.

In sum, early stem hominoids present an interesting mix between anthropoid and hominoid types, allowing for confusion to persist, something we expect during the precursor period.[236] Within the general time frame of 19–15 mya, the entire precursor period, stem apes retain the earlier mobility pattern, a climber that's prone on all fours. The loss of the tail, along with encephalization, is the major hint that a grade level change is coming, however.[237]

In the region of the Major Node (15.65 mya) and its explicit period, we expect greater clarity for the hominoid type and a grade-level change. Yet, as this is the third node in a series, we can also expect a greater degree of invisibility. A fossil gap between 17 and 7 million years has often been noted in the literature, as great ape fossils are extremely rare. Fortunately, recent finds have begun to fill this gap and grant us confirmation.

About 5% after the Major Node, *Nacholapithecus kerorri*, a large bodied hominoid has been dated to 15 mya.[238] Although still an arboreal quadruped like *Proconsul*, a variety of movement has become present—vertical climbing, quadrumanous clambering, suspension, and bridging behaviors.[239] The skeleton exhibits enlarged forelimb joints and longer manual fingers, leading *Nacholapithecus* to be classified as an upright or "orthograde climber."[240] This is the earliest evidence we have for more forelimb-dominating behavior in primates, which is a critical hominoid trait.

A little past the 10% explicit node (14.6 mya), we find *Kenyapithecus wicker*, dated 14 mya.[241] The facial feature of high cheek bones indicates that this hominoid may be more closely related to great apes and humans than to other taxa.[242] Other traits point to the human direction, including thick-enameled teeth, robust jaws with small front teeth, reduced canines, and a shorter face.[243] While almost nothing else is known about the body of *Kenyapithecus*, a joint from the humerus bone shows that this hominoid was semi-terrestrial.[244]

At the 1/4 explicit node (13.16 mya), a recent spectacular fossil helps fill the 10 million year gap. A full cranium of the gibbon-like *Nyanzapithecus alesi* has been dated to 13 mya.[245] The primate's projected weight is 25 pounds, meaning its brain was three times the size of current anthropoids from the prior node, helping to confirm grade level change.[246] Although the skull and small snout look like a gibbon in many respects, the preservation of the inner ear bones reveal that this ape did not specialize as a fast-moving brachiator, like modern-day gibbons. Rather, *Nyanzapithecus* was a more generalized, slower moving primate. Regarding its place with human evolution, these stem hominoids have been described by the discoverers as being close to the origin of extant apes, before humans diverged.

At the 1/3 explicit node (12.4 mya) a great ape, *Anoiapithecus brevirostris*, has been dated 12.4–12.3 mya.[247] The skull possesses a striking reduction in the sloping of its face, giving a modern facial appearance, comparable to the much later *homo* of the human lineage.[248]

At the same time, *Dyropithecus* appears, exhibiting five cusps in their molars with a Y shaped groove, a derived trait linking all hominoids. As for their body/post-cranial anatomy, *Dryopithecus* exhibits the full constellation of characters found in all living hominoids.[249] Researchers have also found great ape features similar to the gorilla, with a downward tilted face, nasal architecture, and upper mouth bones.[250]

A little after the 1/3 explicit node, *Pierolapithecus catalaunicus* from Spain has been dated 12–11.9 mya. This major find presents a clear confirmation of modern ape-like traits, including an orthograde body design and facial morphology consistent with a great ape.[251] The former trait should not be passed over lightly and bears repeating, that this is the first instance of upright trunk adaptations for the great ape.[252] Its cumulative traits, including vertebrate and ribs that resemble gibbons, are evidence for being a basal member of the African ape/human clade.[253] With these various physical traits, especially orthogrady, we are on the cusp of human evolution.

If we extend a little further out than usual, halfway between the 1/3 explicit node and mid-node, *Danuvius*, a recently discovered great ape has been dated to 11.62 mya. The behavioral dynamism and developmental flexibility of hominoids continue with a novel form of positional behavior—"extended limb clambering." *Danuvius* has adaptations of both bipeds and suspensory apes, and so can serve as a model for the common ancestor of great apes and humans.[254]

In sum, the surround of evidence at this node shows the attainment of hominoid status. The precursor period brought evidence for encephalization and loss of a tail, while the suite of traits we attribute to great apes emerges by the explicit period. In the end, a skeletal show of mobility, a sheer increase in brain size, and comparison to existing hominoids confirm a grade level change at the 9th Major Node.

We have traveled far from our first primate of miniature size to the hominoids of more human size. We are now ready for the momentous step that takes us to the first human.

Log Scale of 9th Node Hominoids/Apes

+ Mid-node 11.1

Danuvius /biped & suspensory adaptations 11.62

*1/3 Explicit node (12.4 mya) *Pierolapithecus* 12 First upright/ orthograde hominoid

Divergence of chimpanzees and orangutans 12.5

Dyropithecus 12.4–12.3/full hominoid characters

Anoiapithecus *breviostris*/ great ape 12.4–12.3

*1/4 Explicit node (13.2 mya) *Nyanzapithecus alesi*/gibbon-like cranium 13

*10%   Log   distance   (14.6   mya)   *Kenyapithecus wickeri*/Semi-terrestrial 14

Nacholapithecus kerorri 15

$2^{23.9}$ ◉ 9th **Major Node (15.7 mya) HOMINOIDS** Gorilla divergence 15.1

*10% Log distance (16.7 mya) *Afropithecus* 17.5–17/ Thick enameled teeth

*1/4 Precursor node (18.6 mya) *Proconsul heseloni* 17.9–17.5 *Morotopithecus* 17.5

*1/3 Precursor node (19.7 mya) Definite *Proconsul* 19.6

+ Mid-node (22.1 mya) Oldest *Proconsul* 22 Gibbons diverge 23–16

10th NODE: HUMAN/HOMININ
7.8 MYA

Sahelanthropus

With the tenth node we have the momentous occurrence of the first biped hominid, claiming human status. Recent terminology puts both great apes and humans as hominids, with humans alone being termed hominin. Thus, we are looking for the point when humans separated from the ancestors that produced the great apes. Put another way, it's when hominids became more closely related to humans than to great apes.

Human traits include habitual bipedal locomotion, over-the-top neuralization, and dependence on others making for tight social bonds. Interestingly, change in movement preceded neural expansion, though significant neural reorganization likely occurred earlier, prior to the node. Brain organization has been cited as the critical factor in neural evolution, starting with anthropoids.[255] That is, cerebral morphology can be decoupled from enlarging brain size.[256] Such reorganization would likely accompany novel capacities in hominins, such as precision gripping for complex tasks, complex social interactions that are less aggression based, and expansion into varied terrestrial environments.

However events exactly transpired, the debut of humans is associated with a marked increase in consciousness. Researchers have struggled to state what that means, with hypotheses such as the capacity to symbolize, or the ability to recognize causal effects.[257] Although the timing of this inner dimension will remain largely invisible, we can summon a confluence of evidence regarding bipedality, neuralization, and sociality.

Let us now trace the arc of evolutionary finds and see what is revealed by a logarithmic patterning. As stated earlier, the 10th node of evolution places hominin/human emergence at 7.8 million years ago. We start by looking for finds that show land-based activity. At the 1/3 precursor node (9.86 mya), *Nakalipthecus* from Kenya has been dated at 9.85 mya.[258] Its large size, that of a female gorilla, and dental morphology indicates a partly terrestrial life style. Jaw and tooth traits––thick-enamelled molars, thickened mandular (jaw) body, and low-crowned canines as thick as long––enabled eating of hard foods like seeds and nuts.[259] Modern-day humans have thick and large enameled teeth, which enable access to a variety of food and makes for ecological success.

At the 1/4 precursor node (9.3 mya), *Ouranopithecus* from Greece has been dated at 9.3–9.5 mya.[260] Three traits indicate proximity to the human lineage: (1) Thick enameled teeth, (2) Reduced canines and lack of canine honing complex, and (3) A more vertical profile of the upper face. The thickness of the enamel, the pattern of wear on their teeth, and their paleohabitat, along with this hominid's large size (that of a female gorilla) suggests that *Ouranopithecus* spent considerable time on the ground.[261] Recent finds of finger bones or phalanges further confirm that *Ouranopithecus* was terrestrial, whose motion, however, still matches that of a quadrupedal primate.[262] The relationship of this hominoid to the human line is debated with good arguments on both sides.[263] Dental traits resemble australopithecines, our fossil ancestors, while a recent study places its facial features as closer in proximity to the gorilla than to Homo line.[264] We expect such dispute in the precursor period. Even if *Ouranopithecus* is to the side of the human lineage, its more human-like traits being a convergence, it reveals the evolutionary potential at this time.

As we approach the Major Node, we hope to find a clearer signal for the human type. About 10% prior to the Major Node, *Oreopithecus* inhabited a protected island environment in Italy, dated 8.2 mya. At least four traits point towards the human type:

(1) Skeletal structures indicating upright behaviors including a "platform for postural harvesting" and "biped shuffling."[265] The forward curvature of the lumbar/lower back is human like and implies a form of bipedal locomotion.[266] This trait, however, is not seen as indicating habitual bipedalism.[267]

(2) Pad-to-pad precision gripping by hands, reflecting more complex forelimb activity.[268]

(3) Significant canine reduction and lack of a canine honing complex, indicating a drop in male agonistic behavior.[269]

(4) Fine structures of the inner ear showing closer proximity to the human line than to other hominoids.[270]

(5) Last, this enigmatic ape possessed a relatively vertical face, although a relatively small brain for its body size.[271]

As far as its hominin status, considerable conjecture remains as to the status of *Oreopithecus*, as it "presents a mosaic of dental, cranial and skeletal characters unknown in living primates."[272]

We come to the explicit period, where we look for confirmation for the first human/hominin, and we are not disappointed. A little prior to the 1/4 explicit node (6.6 mya), a remarkable fossil find presents a complete hominid skull that is dated by most researchers to ~7 mya.[273] *Sahelanthropus* has a human-like skull, a relatively flat face, and reduced canines lacking the shearing function found in apes. The lack of an ape-like canine cutting complex is a key trait showing hominin status.[274] Its tooth enamel is of intermediate thickness, more closely approximating the human type.[275]

A recent virtual reconstruction of *Sahelanthropus*'s crushed and heavily distorted skull shows human-like eye sockets, and a rounder, larger skull than originally thought. The cranial capacity has doubled to 370–380 cc, compared to the large dental apes of the prior node.[276] Although brain size is not a relative increase when adjusted for body size, sheer size increase meant more brain matter, and one likely attended by neural reorganization.

There is evidence supporting bipedal behavior with *Sahelanthropus*. The skull shows neck muscles were attached in the same way as for a bipedal hominin. The position of the skull over the body, the angle of the foramen magnum, is closer to humans than apes. Recent study has confirmed that the relationship between the angle of neck opening and bipedality is reliable.[277] Last, recent fossil finds of the first postcranial remains, a left femur and two ulnae, associated with the *Sahelanthropus* site, indicate that this hominin was "already a terrestrial biped," in which careful climbing was also a significant part of its movement.[278]

In short, *Sahelanthropus* was a biped, able to roam the palm, grove-like vegetation zone and mixed grasslands of Chad in Central Africa.[279] Although dissenting opinions are present,[280] *Sahelanthropus* is generally accepted as a basal hominin.[281]

By the 1/3 explicit node (6.2 mya), we hope to find confirmation for grade-level change and hominin status. From the Tugen hills of Kenya, *Orrorin* has been dated securely to 6.14 mya. Rather than a savannah, *Orrorin* roamed humid environments, a variegated forested countryside.[282] This short-faced hominid possessed thick dental enamel, small teeth relative to its body size, and an elongated femur bone, all indicating that Orrorin was a hominin.[283] Of particular importance is the femur, which has a thinner neck to the top of the thighbone, and a groove for a tendon found only in humans.[284] This first direct evidence for bipedality has recently been confirmed by a range of quantifiable measures for a strong anatomical similarity to Homo sapiens.[285] Based on this leg bone, the body weight of a young adult is estimated to be 35–50 kg, or 77–110 pounds.[286] To compliment these fossil finds, a thumb phalanx suggests a precision grip. Although no cranial evidence is available to complete the picture, emerging consensus describes *Orrorin* as a bipedal hominin.[287]

Another line of evidence can be considered as showing hominin confirmation for this node. Phylogenetic analysis and "Bayesian hypothesis testing" can quantify traits to see if they add up to hominin status or not. In a recent thorough study, statistical findings support the dates of our logarithmic scale. From the precursor period, *Ouranopithecus* was placed firmly as a stem member of the African ape and human clade, with mixed support for *Oreopithecus*. With regard to the explicit period, the analysis "strongly supports" *Sahelanthropus* and *Orrorin* as basal hominins.[288]

To summarize, the tenth logarithmic pulse of evolution is characterized by the arrival of the first hominin or human. While the precursor period presented terrestrial behavior and a mix of human and great ape traits, the explicit period offered direct evidence for hominin status and bipedal behavior. Specific traits include a human-like flatter face, sheer increase in brain size, reduction of canines, and the all-important trait of bipedality. A novel mode of locomotion, bringing new experiences and ways of being, no doubt helped promote greater neuralization and consciousness.

Some time ago, Teilhard de Chardin's offered an insightful view, when only a tenth of the fossil evidence was available. For Chardin, the first human resulted from only a "tiny tangential increase" in the brain, bringing about an in-depth change, a boiling over, into "supersensory relationships and representations."[289]

Log Scale of 10th NODE
HOMININS/ HUMANS

*1/3 Explicit node (6.2 mya) *Bipedal Orrorin* 6.14

*1/4 Explicit node (6.6 mya) *Sahelanthropus/bipedal* 1st hominin ~ 7 (6.8–7.2)

2$^{22.9}$ ◙ Major Node (7.8 mya) HOMININS/HUMANS

*10% log distance (8.4 mya) *Oreopithecus/bipedal shuffling* 8.2

*1/4 Precursor node (9.3 mya) *Ouranopithecus* 9.5–9.3 *Dryopithecus* 9.5

*1/3 Precursor node (9.86 mya) *Nakalipithecus nakayamai* 9.9–9.8

CONCLUSION

We have traveled far in our journey from life's origin to the first humans, traversing ten major pulses of evolution. In our exploration, we have found that evolution has pulses that can be mapped with logarithmic regularity. Furthermore, the evidence around each node shows a logarithmic patterning that moves from uncertainty to certainty. We have documented too, that at each Major Node, the qualitative leap occurs with greater consciousness, capacity for movement, and a deepening social connection. This movement to the human type is not to deny there is an equally significant horizontal dimension to evolution. The interweaving of ecological domains with varieties of ecotypes among nature reveals a parallel extravagance in number, variety and emergent novelty.[290]

The journey is not over with these evolutionary pulses. They continue onward for another eight pulses, taking us into the paleolithic era, as humans come to express greater cultural and symbolic activity with art and music. Then, at some point in the Neolithic, the nature of time shifted into historical time with a 15-year generational patterning.[291] Beyond that, the present-day global era is altering the fabric of time yet again. But evolutionary time remains our undercurrent and still drives us to greater consciousness.

What this portends for the future can be considered hopeful. The path of increasing consciousness, mobility, and social connection does not appear to be readily ending. While we can't fully account for the beginning of things, nor their end point, the in-between reveals this amazing logarithmic patterning. It's as if the structure of time itself is unfolding, with as much form as is present in the spatial domain.

Perhaps what is known as divinity or some form of "cosmic consciousness" has a hand in all this, not only at the beginning and the end of things as traditionally envisioned--but in the *in-between*, an amazing patterning with regular pulses of greater consciousness. It seems that the world religions are aware of this journey in time, and how new ages emerge. But here we have brought the fruits of science, and the evolutionary witness for our past, to show that life has been continually gifted with greater freedom to move, greater awareness, and a greater capacity to love one another. We are invited to join in this journey and help build the universe.

TOTAL LOG CHART
10. HOMININS/HUMANS 7.8 Mya

*1/3 Explicit node (6.2 mya) *Bipedal Orrorin* 6.14

*1/4 Explicit node (6.6 mya) *Sahelanthropus/bipedal* 1st hominin 6.8-7.2

$2^{22.9}$ ✪ Major Node (7.8 mya) HOMININS
*10% log distance (8.4 mya) *Oreopithecus/bipedal shuffling* 8.3

*1/4 Precursor node (9.3 mya) *Ouranopithecus* 9.5-9.3 *Dryopithecus* 9.5

*1/3 Precursor node (9.86 mya) *Nakalipthecus nakayamai 9.9-9.8*

9. HOMINOIDS/APES 15.7 Mya

*1/3 Explicit node (12.4 mya) *A. breviostris* 12.4-12.3 *Dyropithecus* 12.4-12.3 *P. catalaunicus* 12-11.9

*1/4 Explicit node (13.2 mya) *Nyanzapithecus alesi*/cranium 13

*10% Log distance (14.6 mya) *Kenyapithecus wickeri*/Semi-terrestrial 14

$2^{23.9}$ ✪ 9th Major Node (15.7 mya) HOMINOIDS
*10% Log distance (16.7 mya) *Afropithecus* 17.5-17/ Thick enamel

*1/4 Precursor node (18.6 mya) *Proconsul heseloni* 17.9 - 17.5 *Morotopithecus* 17.5

*1/3 Precursor node (19.7 mya) Definite *Proconsul* 19.6

+ Mid-node (22.1 mya) Earliest *Proconsul* (23.5)

8. ANTHROPOIDS/MONKEYS 31 Mya

*1/3 Explicit node (24.8 mya) Hominoid/ape divergence? 25-23

*1/4 Explicit node (26.3 mya) *Kamoyapithecus*/ "Dental ape" 24-27.5

$2^{24.9}$ ✿ 8th **Major Node (31.3 mya) Anthropoids** *Aegyptopithecus* 29-30

    *1/4 Precursor node (37.2 mya) *Catopithecus* 35.7

    *1/3 Precursor node (39.4 mya) Marmoset divergence 35.7 - 48.2

    + Mid-node (44.3 mya) *Eosimias* 45 / Basal anthropoid

7. PRIMATES 63 Mya

    * 1/3 Explicit node (49.7 mya) Crown Primate/Tarsiids *Xanthorhysis* 50

    Shoshonius 50

    *1/4 Explicit node (52.7 mya) *Definite* primate/*Teilhadina asiatica* 54.97

$2^{25.9}$ ✿ 7th **MAJOR NODE (62.6 mya) PRIMATES** *Altiatlasisus* 60

    *10% Log distance (67 mya) *Plesiadapiform*s/primtomorphs 66

    *1/4 Precursor node (74.5 mya) Molecular estimates average 71.7

    *1/3 Precursor node (79 mya) Recent molecular estimate 74.4

6. PLACENTAL MAMMALS 125 Mya

    + Mid-node (88.6 mya) 11 types of *Eutherians 90*

    *1/3 Explicit node (99 mya) *Pranyctoides* c. 91.5 *Kulbeckia* 93.9-89.3

    First Eutherian dominated fauna 95

    *1/4 Explicit node (105 mya) *Montanalestes Keebleri* 110

    *10% log distance (117 mya) *Prokennalestes* 115

$2^{26.9}$ ✿ 6th **Major Node (125 Mya) Eutherian** Tree climbing *Eomai* 125

    *1/4 Precursor node (149 mya) *D. leiriensis 150*

    Durlstotherium & Durlstodon 145

    *1/3 Precursor node (158 mya) *Juramai*/stem eutherian 158 mya

    + Mid-node (177 mya) Molecular divergence 180

5. MAMMALS 250 Mya

*1/3 Explicit node (199 mya) *Hadroconium* 195 (Definite mammal)

*1/4 Explicit node (211 mya) *Morganucodon* 196-200 (1st mammal)

*15% Log distance (226 mya) *Adelobasileus* 225 (Possible mammal)

$2^{27.9}$ ✪ 5th **Major Node (250 Mya) Mammal** *Thrinodoxon* 250 Cynodonts 255

*10% Log distance (268 mya) Therapsids/ Expanded range 270

*1/4 Precursor node (298 mya) Sister group of Therapsids 300

*1/3 Precursor node (316 mya) Mammals diverge 315 Amniotes 310

4. VERTEBRATES 501 Mya

+ Mid-node (354 mya) Transitional tetrapod forms 380-325

* 1/3 Explicit node (398 mya) Lobe fin fish/tetrapod 407-409

*1/4 Explicit node (421 mya) Lungfish 412-417

*20% Log distance (436 mya) Gnathostomes/ suite of vertebrate traits 440

*10% Log distance(467mya) Galeaspids/ mineralized endoskeleton 460

$2^{28.9}$ ✪ 4th **Major Node (501 Mya) VERTEBRATES**
Conodonts/ mineralized skeleton 478

*10% Log distance (537 mya) Stem vertebrates 532 Pikaia 530

*15% Log distance (556 mya) *Spriginna* 550 *Kimberella* 555-558

*1/4 Precursor node (596 mya) Fossilized sponge 600

*1/3 Precursor node (631 mya) Embryos 630

Sponge biomarkers 635 Sponge-like animals 650-640

+ Mid-node (+708 mya) Oxygen rise 800-600

3. COMPLEX MULTICELLULARITY 1 Bya

*1/3 Explicit node (795 mya) Collagen deposits >779 Metazoan genes 800

*1/4 Explicit node (842 mya) *Protoarenicola* and *Pararenicola*

*10% Log distance (934 mya) Megasphareomorphs 900 Spores 900

$2^{29.9}$ ✸ 3rd Major Node (1 Bya) COMPLEX MULTICELLULAR HOLOZOA 1 bya

*10% Log distance (1.07 bya) Eucaryotic burst 1.1- .9 *Bangiomorpha* 1.047

*1/4 Precursor node (1.19 bya) Rapid diversification eucaryotes 1.2- 1 bya

*1/3 Precursor node (1.26 bya) Molecular est. for last common ancestor of Opisthokonta 1.389-1.240

2. EUCARYOTIC CELLS 2 Bya

*1/3 Explicit node (1.59 bya) Undisputed eukaryotes *T. plana* 1.6

*1/4 Explicit node (1.68 bya) Large acritarchs (1.7 -1.8) Multicelled 1.7

* 15% log distance (1.8) Spheroidal phytoplankton 1.8

*10% Log distance (1.87 bya) *Grypania* 1.85

$2^{30.9}$ ✸ 2nd Major Node (2 Bya) EUCARYOTES *Oscillatoriopsis* 4

*10% Log distance (2.15 bya) Definite presence of Cyanobacteria 2.15

*1/4 Precursor node (2.38 bya) Great Oxygen event 2.32 Biomarkers 2.4

*1/3 Precursor node: (2.52 bya) Molecular estimates 2.5 - 2

1. ORIGIN OF LIFE 4 Bya

*1/3 Explicit node (3.18 bya) Sulfide metabolism 3.235
Submarine life 3.48-3.22
*1/4 Explicit node (3.37 bya) Stromatolites 3.4 Apex Filaments 3.456
Methane microbes 3.5
*1/10 node (3.73 bya) Isotopes indicating life 3.85
Putative fossils at hydrothermal vents 3.77
$2^{31.9}$ ✪ 1st **Major Node (4 Bya) LIFE** Molecular estimates 4.0
Light Carbon from Canada 4
5% node (4.15 bya) Light carbon from Greenland 4.15
*1/10 node (4.3 bya) Water 4.3-4.4 Oldest rocks 4.28
Light carbon from Australia 4.25
*1/4 Precursor node (4.77 bya) Solar system/Earth forms 4.57/ 4.54 Crust forms 4.5
*1/3 Precursor node (5.05 bya) Proto-sun forms 5.05

GLOSSARY

Allometric: Size and shape changes in an organism, in which given structures grow at different rates than others. Thus the growth rate of a baby's head is much greater in the first year than the legs of its body.

Anthropoid: A suborder of primates represented by extant monkeys

Arboreal: A tree-dwelling animal, often used to describe primates.

Bayesian: An interpretation of the concept of probability. Instead of the frequency or propensity of a phenomenon, probability is interpreted as a reasonable expectation representing your current state of knowledge.

Clade: A group of organisms believed to have evolved from a common ancestor.

Crown Group: In the line leading to a type of organism, fossils which can be identified as basically the same as living or extant members.

Dimorphic: Having two different body shapes. For example, males and females of animals have different size ranges and characteristics.

Eukaryotic: Cells, much larger and complex than prokaryotes. which have internal organelles, such as mitochondria which vastly increases energy production, and a nucleus which harbors the organism's DNA.

Eutherians: Another word for placental mammals.

Encephalization Quotient: A measure of the relative size of the brain of a particular species compared with the expected value for members of the group to which it belongs. Modern humans have an EQ of 6, meaning that their brain mass is six times greater than a typical mammal.

Grade Level Change: A marked qualitative change, readily identifiable in the spheres of neural enhancement, mobility, and social capacity.

Hominoid: A primate suborder which includes the great apes, such as chimpanzees and gorillas.

Last Common Ancestor: This first ancestor of the crown group, fossils which are close enough to be identified with current, living survivors.

Logarithmic: When units keep expanding or subtracting in exponential fashion, using bases and exponentials. For example, using two as a base, you have larger and larger increases as the exponential or logarithm goes up by one value. Thus, the logarithmic increases: $2^2 = 4$; $2^3 = 8$; $2^4 = 64$; $2^5 = 128$ and so on.

Major Node: This is a point in time where major grade level change is hypothesized in evolution, such as the advent of the first vertebrate or first mammal.

Minor Node: This are points in time, where the evidence for grade level change is assessed, typically at ¼ and 1/3 log distance before and after a Major Node.

Complex Multicellularity: Organisms which have more than one cell type and are interdependent, or depend on one another to survive.

Explicit Nodes: Points in time when the evidence for grade-level change has become clear.

Precursor Nodes: Points in time when the evidence for grade-level change is approximate and uncertain.

Placental Mammal: A mammal which carries its young internally and bears their live young. Other traits include direct interconnecting of the hemispheres of the brain by the corpus callosum.

Procaryotic: The earliest form of life, characterized by cells, which have a semipermeable membrane, the ability to find energy sources, metabolize and sustain its complexity, as well as to reproduce.

Stem group: Early forms of a given type of organism, which have only some of the traits, many of which are evolutionary dead ends.

Ur-Metazoan: The hypothetical last common ancestor of all animals.

LIST OF ILLUSTRATIONS

ENDNOTES

[1]. Martin, R. et al., 2007. Primate origins: Implications of a Cretaceous Ancestry. *Folia Primatology* 78:277-296.

[2]. Zak, M., 2005. From reversible thermodynamics to life. *Chaos, Solitons and Fractals* 26: 1019-1033.

[3]. Stebbing A. 2016. The Unity of Opposites and the 2nd Law of Thermodynamics, with a biological outcome. *Syntropy* 1: 1-19.

[4]. Pierce, B. et al., 2018. Constraining the Time Interval for the Origin of Life on Earth. *Astrobiology* 18 (3): 343-264.

[5]. Cavosie, A., Valley, J., Wilde, S., 2005. Magmatic δ^{18} O in 4400-3900 Ma detrital zircons: A record of alteration and recycling of

crust in the Early Archaean. *Earth and Planetary Science Letters* 235: 663-681.

[6]. Nemchin et al., 2008. A light carbon reservoir recorded in zircon-hosted diamond from the Jack Hills. *Nature* 454: 92-95.

[7]. Holland, H., 1997. Evidence for life on earth more than 3850 million years ago. *Science* 275(3): 38-39.

[8]. Bell, et al., 2015. Potentially biogenic carbon preserved in a 4.1 billion-year-old zircon. *Proceedings of the National Academy of Sciences*, USA 112, 14518-14521.

[9]. Javaux, E., 2019. Challenges in evidencing the earliest traces of life. *Nature* 752:451-460.

[10]. Battistuzzi, F. et al. 2004. A genomic analysis of prokaryote evolution: insights into the origin of methanogenesis, phototrophy, and the colonization of land. *BMC Evolutionary Biology* 4:44.

[11].Tashiro, T. et al. 2017. Early trace of life from 3.95 Ga sedimentary rocks in Labrador, Canada. Nature 549(28): 516-518.

[12]. Nutman, et al., 2010. 3700 Ma pre-metamorphic dolomite formed by microbial in the Isua supracrustal belt (W. Greenland). Simple evidence for life? *Precambrian Research* 183: 725-737.

[13]. Yoshiya, K., 2015. In-situ iron isotope analysis of pyrites in ~3.7 Ga sedimentary protoliths from the Isua supracrustal belt, southern West Greenland. *Chemical Geology* 401: 126-139.

[14]. Dodd, et al., 2017. Evidence for early life in Earth's oldest hydrothermal vent precipitates. *Nature* 543:60-64.

[15]. Hall, B., & Hallgrimsson, B. (2007). *Strickberger's Evolution, 4th edition* (p. 145). Boston: Jones and Bartlett.

[16]. Ueno, Y. et al., 2006. Evidence from fluid inclusions for microbial methanogenesis in the early Archaean era. *Nature* 440 526-519.

[17]. Schopf, W. et al., 2018. SIMS analyses of the oldest known assemblage of microfossils document their taxon-correlated carbon isotope compositions. *Proceedings of the National Academy of Sciences* 115 (1) 53-58.

[18]. Fumes, H., Banerjee, N., Staudigel, H., & de Wit, M., 2004. Early life recorded in Archean pillow lavas. *Science* 304: 578-581.

[19] Homann, M. 2019 Earliest life on Earth: Evidence from the Barberton Greenstone Belt, South Africa. *Earth-Science Reviews* 196: 102888.

[20]. Rice, P., & Moloney, N., 2008. *Biological Anthropology and Prehistory, 2nd edition*. Boston, Mass: Pearson.

[21] **Gaucher, C. &I Frei, R., 2019. The Archean-Proterozoic Boundary and the Great Oxidation Event. In (ed. Sial, A.) *Chemostratigraphy Across Major Chronological Boundaries*. John Wiley & Sons Inc.**

[22] Hodges, S. et al., 2004. A molecular timescale of eukaryote evolution and the rise of complex multicellular life. *BioMed Central*.

[23]. Rasmussen, B. et al., 2008. Reassessing the first appearance of eukaryotes and cyanobacteria. *Nature* 455, 1101-1104.

[24] Sergeev, V., 2018. The Biostratigraphic Paradox of Precambrian Cyanobacteria: Distinguishing the Succession of Microfossil Assemblages and Evolutionary Changes Observed among Proterozoic Prokaryotic Microorganisms. *Paleontol. J.* 52: 1148–1161.

[25] Rozanov, A. &. Astafieva, M. 2008. Prasinophyceae (Green Algae) from the Lower Proterozoic of the Kola Peninsula. *Paleontological Journal* 42(4): 425–430.

[26]. Yin, et al., 2020. Microfossils from the Paleoproterozoic Hutuo Group, Shnxi, North China: Early evidence for eukaryotic metabolism. *Precambrian Research* https://doi.org/10.1016/j.precamres.2020.105650.

[27] Hofmann, H. 1976. Precambrian Microflora, Belcher Islands, Canada: Significance and Systematics. *Journal of Paleontology* 50(6): 1040-1073.

[28] Tappan, H. 1976. Possible eukaryotic algae (Bangiophycidae) among early Proterozoic microfossils. *GSA Bulletin* 87 (4): 633–639.

[29] Sergeev, V., 2018. The Biostratigraphic Paradox of Precambrian Cyanobacteria: Distinguishing the Succession of Microfossil Assemblages and Evolutionary Changes Observed among Proterozoic Prokaryotic Microorganisms. *Paleontol. J.* 52: 1148–1161.

[30]. Knoll A. et al., 2006. Eukaryotic organisms in Proterozoic oceans. *Philosophical Transactions of the Royal Society* 361: 1023-1038.

[31]. Schopf, J., 1999. *Cradle of Life: Discovery of Earth's earliest fossils* (p. 243). Princeton: Princeton University Press.

[32] Betts, H. et al. 2018. Integrated genomic and fossil evidence illuminates life's early evolution and eukaryote origin. *Nature Ecology & Evolution* 3: 338-344.

[33]. Teyssedre, B., 2006. Are the green algae (phylum Viridplantae) two billion years old? Carnets de Geologie/Notebooks on Geology Article 2006/03 (CG2006_A03)

[34] Pang, K. et al. The nature and origin of nucleus-like intracellular inclusions in Paleoproterozoic eukaryote microfossils, Geobiology 11 499-510.

[35]. Qu, Y. et al., 2018. Carbonaceous biosignatures of the earliest putative macroscopic multicellular eukaryotes from 1630 Ma Tyuanshanzi Formation, north China. *Precambrian Research* 304: 99-109.

[36]. Javaux, E. & Lepot, K., 2018. The Paleoproterozoic fossil record: Implications for the evolution of the biosphere during Earth's middle age. *Earth-Science Reviews* 176: 68-86.

[37]. Knoll, A. Javauz, E. Hewitt, D. & Cohen, P., 2006. Eukaryotic organisms in Proterozoic oceans. *Philosophic Transactions of the Royal Society* 361, 1023-1038.

[38]. Shi, M. et al., 2017. An eukaryote-bearing microbiota from the early mesoproterozoic Gaoyuzhuang Formation, Tianjin, China and its significance. *Precambrian Research* 303: 709-726.

[39]. Zhu et al., 2016. Decimetre-scale multicellular eukaryotes from the 1.56 billion year old Gaoyuzhuang formation in North China. *Nature Communications*, DOI: 10:1038/ncomms11500.

[40] Hedges, S. et al., 2015. Tree of life reveals clock-like speciation and diversification. *Mol. Biol. Evol.* 32(4): 835-845.

[41]. Butterfield, N. 2009. Modes of pre-Ediacaran multicellularity. *Precambrian Research* 173: 201-211.

[42]. Grau-Bove, X et al. 2017. Dynamics of genomic innovation in the unicellular ancestry of animals. *Elife*; Research Article 6:e26036. Doi: 10:7554/eLIfe.26036.

[43]. Richter, D., 2018 Gene family innovation, conservation and loss on the animal stem lineage. *eLife* 2018;7:e34226 DOI: 10.7554/eLife.34226.

[44] Thattai, M. 2019. How contraction has shaped evolution. *eLIfe*: 8: e52805.

[45]. Ruiz-Trillo, I. 2016. What are the Genomes of Premetazoan lineages telling Us about the Origin of Metazoa? (p. 171-194). *In Multicellularity: Origins and Evolution.* Ed. Nikias, K & Newman, S. MIT Press: Cambridge Mass.

[46]. Paps, J. & Holland W., 2018. Reconstruction of the ancestral metazoan genome reveals an increase in genomic novelty. *Nature Communications* 9:1730.

[47]. Parfrey, L. et al., 2011. Estimating the timing of early eukaryotic diversification with multigene molecular clocks. *PNAS*, 108(33): 13624-13629.

[48]. Cunningham, J. et al., 2017. The origin of animals: Can molecular clocks and the fossil record be reconciled? *Bioessays* 39:1 1600120.

[49]. Schopf, J., 1999. *Cradle of Life: Discovery of Earth's earliest fossils.* (p. 254). Princeton University Press.

[50]. Gibson, T. et al., 2017. Precise age of *Bangiomorpha pubescens* dates the origin of eukaryotic photosynthesis. *Geology,* 46(2): 135-138.

[51]. Butterfield, N., 2000. *Bangiomorpha pubescens* n. gen.,n.sp.: Implications for the evolution of sex, multicellularity and the Mesoproterozoic/Neoproterozoic radiation of eukaryotic. *Paleobiology* 26(3): 386-404.

[52]. Tang, Q et al., 2020. A one-billion-year-old multicellular chlorophyte. *Nature Ecology & Evolution* 4: 543B549.

[53]. Loron, C. et al., 2019. Early fungi from the Proterozoic era in Arctic Canada. *Nature* 270: 232-235.

[54]. Dohrman, M. & Worheide, G., 2017. Dating early animal evolution using phylogenomic data. *Scientific Reports,* 7:3599.

[55]. Loron, C., et al. 2019. Organic-walled microfossils from the late Mesoproterozoic to early Neoproterozoic lower Shaler Supergroup (Arctic Canada): Diversity and biostratigraphic significance. *Precambrian Research* 321: 349-374.

[56]. Butterfield, N. 2009. Modes of pre-Ediacaran multicellularity. *Precambrian Research* 173: 201-211.

[57]. Paps, J. & Holland W., 2018. Reconstruction of the ancestral metazoan genome reveals an increase in genomic novelty. *Nature Communications* 9:1730.

[58] Saunders, M. et al. 2014. Electron microscopy reveals unique microfossil preservation in 1 billion-year-old lakes. Journal of Physics: Conference Series 522: 012024.

[59]. Strother, P, Wacey, D. & Wellman C. 2016. An early multicellular holozoan from the 1 Ga Torrdion Group, Scotland. Abstract from *35th International Geological Congress (IGC)* Cape Town, South Africa. Paper Number: 3605. https://www.americangeosciences.org/igc/15511.

[60] Strother, P. Wacey, D., Brasier, M, & Wellman C., 2016. Reconstructed life cycle of a Proterozoic holozoan. Poster Abstract from the *60th Annual Meeting of the Paleontological Association* at the Universite' Claude, Lyon France.

[61] Strother, P. Brasier, M, Wacey D., Timpe L. Saunders, M. & Wellman, C., 2021. A possible billion-year-old holozoan with differentiated multicellularity. *Current Biology* 31: 2658-2665.

[62]. Tikhonenkov, D. et al., 2019. Insights into the origin of metazoan multicellularity from predatory unicellular relatives of animals. *BioR*xiv doi: https://doi.org/10.1101/817874.

[63]. Nielsen, C., 2012. *Animal Evolution: Interrelationships of the Living Phyla, 3rd edition.* (p.7). Oxford University Press.

[64] Brazda V. et al. 2020. The changes in the p53 protein across the animal kingdom pointing to its involvement in longevity. bioRxiv 2020.05.06.080200; doi: https://doi.org/10.1101/ 2020.05.06.080200.

[65] Turner, E.C. 2021. Possible poriferan body fossils in early Neoproterozoic microbial reefs. *Nature.* https://doi.org/10.1038/ s41586-021-03773-z

[66]. Dong, L. Et al., 2008. Restudy of the worm-like carbonaceous compression fossils Protoarenicola, Pararenicola, and Sinosabellidities from early Neoprotereozoic successions in North China, *Palaeogeography, Paleoclimatology, Palaeoecology* 258: 138-161.

[67]. Laflmaee, M. Et al., 2004. Niroinetruc analysis of the Edicarian frond Charniodicus from the Mistaken Point Formation, Newfoundland. *Journal of Palenotology* 78(5): 827-837.

[68]. Neuweiler F, Turner E, & Burdige D., 2009. Early Neoproterozoic origin of the metazoan clade recorded in carbonate rock texture. *Geology* 37: 475-478.

[69]. Brocks, et al., 2017. The rise of algae in Cryogenian oceans and the emergence of animals. *Nature* 548, 578-581.

[70]. Green, S. et al., 2015. Evolution of vertebrates as viewed from the crest. *Nature* 250: 474- 482.

[71]. Maloof, A., 2010. Possible animal-body fossils in pre-Marinoan limestones from South Australia. *Nature Geoscience* 3: 653-659.

[72]. Love et al., 2009. Fossil steroids record the appearance of Demospongiae during the Cryogenian period. *Nature* 457, 718-721.

[73]. Budd, G., 2009. The earliest fossil record of the animals and its significance. In *Animal Evolution: Genomes, Fossils, and Trees,* (ed. Telford M., & Littlewood , D.) Oxford: Oxford University Press.

[74]. Yin, Z, et al., 2019. The Early Ediacaran Caveasphaera Foreshadows the Evolutionary Origin of Animal-like Embryology. *Cell Press,* 29:1-8.

[75]. Ligrone, R., 2019. *Biological Innovations that Built the World: A Four-billion-year Journey through Life and Earth History.* (see pages 331-33) Springer Nature Switzerland.

[76]. Yin, Z. Et al., 2014. Sponge grade body fossil with cellular resolution dating 60 mya before the Cambrian. *PNAS,* https://doi/ 10.1073/pnas 1414677112.

[77]. Bobrovskiy, I. et al., 2018. Ancient steroids establish the Ediacaran fossil Dickinsonia as one of the earliest animals. *Science* 361: 1246-1249.

[78]. Fedonkin, M. et al., 2007. New data on Kimberella, the Vendian mollusc-like organism (White sea region, Russia): palaeoecological and evolutionary implications in Vickers-Rich, Patricia; Komarower, Patricia, *The Rise and Fall of the Ediacaran Biota.*

[79]. Xiao, S., & Laflamme, M., 2008. On the eve of animal radiation: phylogeny, ecology and evolution of the Edicara biota. *Trends in Ecology and Evolution* 24(1): 31-40.

[80]. Morris, S., 2012. Pikaia gracilens Walcott, a stem-group chordate from the Middle Cambrian of British Columbia. *Cambridge Philosophical Society;* https://doi.org/10.1111/ j.1469-185X.2012.00220.x.

[81]. Ligrone, R., 2019. *Biological Innovations that Built the World: A Four-billion-year Journey through Life and Earth History.* (see pages 334) Springer Nature Switzerland.

[82]. Janvier, P., 2015. Facts and fancies about early fossil chordates and vertebrates. *Nature* 520: 483-489.

[83]. Benton, M., 2015. *Vertebrate Palenotology*, 4th edition. (p. 46.). Wiley Blackwell Publishing.

[84]. Shu, D. et al., 2003. Head and backbone of the Early Cambrian vertebrate *Haikouichthys*. *Nature* 421: 526-529.

[85]. V. V. Missarzhevsky., 1973. Conodont-shaped organisms from the Precambrian-Cambrian boundary beds of the Siberian Platform and Kazakhstan]. Trudy Instituta Geologii I Geofiziki SO AN SSSR (Konodontoobraznye organizmy iz pogranichnykh sloev kembriya I dokembriya Sibirskoj platformy i Kazakhstana. Problemy paleontologii i biostratigrafii nizhnego kembriya Sibiri i Dalínego vostoka 53-57.

[86]. Murdock, et al., 2013. The origin of conodonts and of vertebrate mineralized skeletons. *Nature* 502: 546-549.

[87]. Morris, S. & Caron, J., 2014. A primitive fish from the Cambrian of North America. *Nature* 512: 419B422.

[88]. Kardong, K., 2019. *Vertebrates: Comparative Anatomy, Function, Evolution, 8th edition.* (p. 80) New York: McGraw Hill Education

[89]. Long, J., 2011. *The Rise of Fishes: 500 Million Years of Evolution.* (p. 33). Baltimore: Johns Hopkins University Press.

[90]. Zhong, W. et al., 2005. Histology of the galeaspid dermoskeleton and endoskeleton, and the origin and early evolution of the vertebrate cranial endoskeleton. *Journal of Vertebrate Paleontology* 25(4): 745-756.

[91]. Gai Z. et al., 2011. Fossil jawless fish from China foreshadows early jawed vertebrate anatomy. *Nature* 476:324-27.

[92]. Tinn, O/ & Märss T., 2018. The earliest osteostracan Kalanaspis delectabilis gen. et sp. nov. from the mid-Aeronian (mid-Llandovery, lower silurian) of Estonia. *Journal of Vertebrate Paleontology*, DOI: 10.1080/02724634.2017.1425212.

[93]. O'Shea, J., 2019. The dermal skeleton of the jawless vertebrate Tremataspis mammillata (Osteostraci, stem Gnathostomata). *Journal of Morphology* 280:999-1025.

[94]. Afanassieva, O., 2016. On the Growth and Regeneration of the Exoskeleton in Early Jawless Vertebrates (Osteostraci, Agnatha). *Doklady Biological Sciences* 466: 32B35.

[95]. Benton, M., 2015. *Vertebrate Palenotology, 4th edition*. (p. 56-57). Wiley Blackwell Publishing.

[96]. Donoghue, P., & Purnell, M., 2005. Genome duplication, extinction and vertebrate evolution. *Trends in Ecology and Evolution* 20(6): 312-319.

[97]. Striedters, G & Northcutt, R., 2020. *Brains Through Time: A Natural History of Vertebrates*. (p. 140) Oxford University Press.

[98]. Wang, J., 1991. The antiarchi from the early Silurian of Hunan. *Vertebrata PalAsiatica* 29: 240-244.

[99]. Benton, M. et al., 2020. *Cowen's History of Life, 6th edition*. (p. 101-102). Wiley Blackwell.

[100]. Zhu, M., Zhao, W., Jia, L. et al., 2009. The oldest articulated osteichthyan reveals mosaic gnathostome characters. *Nature* 458, 469B474. https://doi.org/10.1038/nature07855.

[101]. Benton, M. et al., 2020. Cowen's History of Life, 6th edition. (p. 106). Wiley Blackwell.

[102]. George, D. & Blieck, A., 2011. Rise of the Earliest Tetrapods: An Early Devonian Origin from Marine Environmene *PLoS One*. 2011; 6(7): e22136.

[103]. Beznosov, P, Clack, J. et al., 2019. Morphology of the earliest reconstruct able tetrapod Parmastega aelidae. *Nature* 574: 527-531.

[104]. Clack, J., 2002. Patterns and Processes in the early evolution of the tetrapod ear. *Journal of Neurobiology* 53: 251-264.

[105]. Lovegrove, B., 2016. A phenology of the evolution of endothermy in birds and mammals. *Biological Reviews* Doi: 10.1111/brv.12280.

[106]. Poelmann, R. & Groot, A. 2019. Development and evolution of hte metazoan heart. *Developmental Dynamics* 248: 634-656.

[107]. Oftedal, O., 2002. The mammary gland and its origin during synapsid evolution. *Journal of Mammary Gland Biology and Neoplasia* 7:3 225-252.

[108]. Feldhammer G. et al., 2020. *Mammalogy: Adaptation, Diversity, Ecology, 5th edition*. (see Chapter 4.) Baltimore: Johns Hopkins University Press.

[109]. Spindler, F., 2019. The skull of Tetraceratops insignis (Synapsida, Sphenacodontia). *Paleo Vertebrata* doi:10.18563/pov.43.1.el.

[110]. Kemp, T., 2012. The Origin and Radiation of Therapsids. In *Forerunners of mammals: Radiation, Histology, Biology* Chinsamy-Turan, A. (Ed.) p. 6.

[111]. Lungmus, J. & Angielczyk, K., 2019. Antiquity of forelimb ecomorphological diversity in the mammalian stem lineage (Synapsida). *PNAS* 14: 6903-6907; https://doi.org/10.1073/pnas.1802543116.

[112]. Jones, K. et al. 2018. Fossils reveal the complex evolutionary history of the mammalian regionalized spine. *Science* 361(6408): 1249-1252.

[113]. Benoit, J et al., 2017. Endocranial Casts of Pre-Mammalian Therapsids Reveal an Unexpected Neurological Diversity at the Deep Evolutionary Root of Mammals. *Brain, Behavior, and Evolution* 90(4): 311-333.

[114] Laab, M. & Kaestner. 2017. Evidence for convergent evolution of a neocortex-like structure in a late Permian therapsid. Journal of Morphology 278(8): 1033-1057.

[115]. Benton, M. et al. 2020. *Cowen's History of Life, 6th edition*. (p. 144-45) Wiley Blackwell.

[116]. Olivier, C. et al. 2017. First palaeohistological inference of resting metabolic rate in an extinct synapsid, Moghreberia nmachouensis (Therapsida: Anomodontia). Biological Journal of the Linnean Society, Volume 121, Issue 2, 1: 409B419, https://doi.org/10.1093/biolinnean/blw044.

[117]. Rey, K. et al. 2017. Oxygen isotopes suggest elevated thermometabolism within multiple Permo-Triassic therapsid clades. *Elife*, https://doi.org/10.7554/eLife.28589.001.

[118]. Crompton, A. et al., 2017. Structure of the nasal region of non-mammalian cynodonts and mammaliaformes: Speculations on the evolution of mammalian endothermy. *Journal of Vertebrate Paleontology* DOI: 10.1080/02724634.2017.1269116.

[119]. Benoit, J. 2016. Cranial bosses of Choerosaurus dejageri (Therapsida, Therocephalia): earliest evidence of cranial display structures in eutheriodonts. *PLOS one* E 11(8): e0161457.

doi:10.1371/ journal.pone.0161457

[120]. Benton, M. et al. 2020. *Cowen's History of Life, 6th edition*. (p. 266.) Wiley Blackwell

[121]. Rubidge, B., Sidor, C., 2001. Evolutionary patterns among Permo-Triassic therapsids. *Annual Review of Ecological Systematics* 32: 449-480.

[122]. Benton, M. et al., 2020. *Cowen's History of Life, 6th edition*. (see chart, p. 265.) Wiley Blackwell.

[123]. Rowe, et al., 2011. Fossil Evidence on Origin of the Mammalian Brain. *Science* 322: 955-957.

[124]. Quiroga, J., 1979. The inner ear of two cynodonts (reptilia-therapsida) and some comments on the evolution of the inner

ear from polycosaurs to mammals. *Morphol. (Gegenbaurs)* JB 125(2): 178-190.

[125]. Ruta, M et al., 2013. The radiation of cynodonts and the ground plan of mammalian morphological diversity. *Proceedings of the Royal Society B* 280:20131865.

[126]. Kammerer, C, 2016. A new taxon of cynodont from the Tropidostoma Assemblage Zone (upper Permian) of South Africa, and the early evolution of Cynodontia. *Papers in Paleontology* 2: 3, 387-397.

[127]. Benton, M., 2015. *Vertebrate Palaeontology, 4th edition*. (p. 326). Wiley Blackwell Publishing.

[128]. Benton, M., 2015. *Vertebrate Palaeontology, 4th edition*. (p. 322). Wiley Blackwell Publishing.

[129]. Lautenschlager, S. et al., 2018. The role of miniaturization in the evolution of the mammalian jaw and middle ear. *Proc. R. Soc. B* 285: 20181792. http://dx.doi.org/10.1098/rspb.2018.1792.

[130]. Jasinoski, S. & Abdala, F., 2017. Aggregations and parental care in the Early Triassic basal cynodonts Galesaurus planiceps and Thrinaxodon liorhinus. *PubMed* 28097072.

[131]. Benton, M., 2015. *Vertebrate Palenotology, 4th edition*. (p. 328). Wiley Blackwell Publishing.

[132]. Luo, Zhexi, et al. 1995. Evolutionary Origins of The Mammalian Promontorium and Cochlea. *Journal of Vertebrate Paleontology* 15(1): 113-121.

[133]. Benton, M. et al. 2020. *Cowen's History of Life, 6th edition*. (p. 270) Wiley Blackwell.

[134]. Feldhammer G. et al. 2020. *Mammalogy: Adaptation, Diversity, Ecology, 5th edition*. (see Chapter 4.) Baltimore: Johns Hopkins University Press.

[135]. Rowe, et al., 2011. Fossil evidence on the origin of the mammalian brain. *Science*, 332: 955-957.

[136]. Rowe, T & Shepherd G., 2015. Role of ortho retronasal olfaction in mammalian cortical evolution. *Journal of Comparative Neurology* 524:471B495.

[137]. Rodrigues, P. et al., 2013. Digital Reconstruction of the Otic Region and Inner Ear
of the Non-Mammalian Cynodont Brasilitherium riograndensis (Late Triassic, Brazil) and Its Relevance to the Evolution of the Mammalian Ear. *Journal of Mammalian Evolution* 20: 291B307.

[138].O'Meara, R. & Asher, R., 2016. The evolution of growth patterns in mammalian versus non-mammalian cynodonts. *Paleobiology* 42(3): 439-464.

[139]. Lautenschlager, S. et al., 2017. Morphological Evolution of the mammalian jaw abductor complex. *Biological Review*, 92: 1910-1940.

[140]. Lou, Z. et al., 2001. A new mammaliaform from the early Jurassic and evolution of mammalian characteristics. Science 292 (5521): 1535-1540.

[141]. Lou, Z. et al., 2001. A new mammaliaform from the early Jurassic and evolution of mammalian characteristics. *Science* 292 (5521): 1535-1540.

[142]. Rowe, T & Shepherd G. 2015. Role of ortho retronasal olfaction in mammalian cortical evolution. *Journal of Comparative Neurology* 524:471B495.

[143]. Feldhammer G. et al. 2020. *Mammalogy: Adaptation, Diversity, Ecology, 5th edition.* (see Chapter 4.) Baltimore: Johns Hopkins University Press.

[144]. Hutson, J.D., Hutson, K.N. 2017. An Investigation of the Locomotor Function of Therian Forearm Pronation Provides Renewed Support for an Arboreal, Chameleon-like Evolutionary Stage. *Journal of Mammal Evolution* 24: 159-177.

[145]. Manley, G., 2012. Evolutionary Paths to Mammalian Cochleae. *Journal of the Association for Research in Otolaryngology* 13:733-745.

[146] Kaas, J. 2020. The Organization of the Neocortex in Early Mammals. In Kass, J. ed. *Evolutionary Neuroscience.* (pp. 339-341) London: Academic Press.

[147]. Mihrshahi, R. 2005. The Corpus Callosum as an Evolutionary Innovation. *Journal of Experimental Zoology* 306B: 8-17.

[148]. Ginsburg, S. & Jablonka, E., 2019. The Evolution of the Sensitive Soul: Learning and the Origins of Consciousness. Cambridge, MA. Mass Institute of Technology.

[149]. Lou, Z. et al. 2001. Dual origin of tribosphenic mammals. *Nature* 409: 53-57.

[150]. Lou, Z. et al., 2011. A Jurassic eutherian mammal and divergence of marsupials and placentals. *Nature* 476: 442-445.

[151]. Meng, J., 2014. Mesozoic mammals of China: implications for phylogeny and early evolution of mammals. *National Science Review* 1: 521-542.

[152]. Carlton, R. 2019. *A Concise Dictionary of Palenotology, 2nd Edition.* (p. 9) Springer Nature Switzerland.

[153]. Meng, et al. 2015. An arboreal docodont from the Jurassic and mammaliaform ecological diversification. *Science* 13: 764-768.

[154]. Vázquez-Molinero R. et al., 2001. Comparative anatomical investigations of the postcranial skeleton of Henkelotherium guimarotae Krebs, 1991 (Eupantotheria, Mammalia) and their implications for its locomotion. *Mitt. Mus. Nat.kd. Berl., Zool. Reihe* 77(2): 207-216.

[155]. Luo, Z. et al., 2011. Fossil evidence on evolution of inner ear cochlea in Jurassic mammals. *Proceedings of the Royal Society,* 278:28-34.

[156]. Luo, et al., 2012.The petrosal and inner ear of the Late Jurassic cladotherian mammal Dryolestes leiriensis and implications for ear evolution in therian mammals, *Zoological Journal of the Linnean Society* 166 (2): 433B463.

[157]. Sweetman, S., 2017. Highly derived eutherian mammals from the earliest Cretaceous of southern Britain. Acta *Paleontological Polonica* 62(4): 657-665.

[158]. Chen, M., Luo, Z., 2013. Postcranial Skeleton of the Cretaceous Mammal *Akidolestes cifellii* and Its Locomotor Adaptations. *Journal of Mammalian Evolution* 20, 159B189.

[159]. Manley, G., 2012. Evolutionary Paths to Mammalian Cochleae. *Journal of the Association for Research in Otolaryngology.* 13:733-745.

[160]. Ji, Q., 2002. The earliest known eutherian mammal. *Nature,* 416: 816-822.

[161]. Zachos, F. & Asher, R. (eds.), 2018. *Mammalian Evolution, Diversity and Systematics.* (p. 275). In *Handbook of Zoology.* Berlin: de Gruyter

[162]. O'Leary, et al., 2013. The placental Mammal Ancestor and the Post-K-Pg Radiation of Placentals. *Science* 339: 662-667.

[163]. Averianov, A. & Archibald, J., 2015 Evolutionary transition of dental formula in Late

Cretaceous eutherian mammals. *Sci Nat* 102:56

[164]. Bi, S. et al., 2018. An early Cretaceous eutherian and the placental-marsupial dichotomy. *Nature* https://doi.org/10.1038/s41586-018-0210-3.

[165]. Aboitiz, F & Montiel, J., 2015. Olfaction, navigation, and the origin of the isocortex. *Frontiers in Neuroscience* 9 (Article 402).

[166]. Carlton, R., 2019. *A Concise Dictionary of Palenotology,* 2nd *Edition.* (p. 345) Springer Nature Switzerland.

[167]. Lopatin, V. & Averianov. A., 2017. The stem placental mammal *Prokennalestes* from the Early Cretaceous of Mongolia. *Paleontological Journal* 51: 1293B1374.

[168]. Lopatin, A. & Averianov. A., 2017. The Stem Placental Mammal *Prokennalestes* from the Early Cretaceous of Mongolia. *Paleontological Journal* 51(12): 1293-1374.

[169]. Wible, J. et al., 2001. Earliest eutherian ear region: a petrosal referred to *Prokennalestes* from the early Cretaceous of Mongolia. *American Museum Novitates* 3322: 1-44.

[170]. Luo, Z. et al., 2011. Fossil evidence on evolution of inner ear cochlea in Jurassic mammals. *Proceedings of the Royal Society* 278: 28-34.

[171]. Manley, G., 2012. Evolutionary Paths to Mammalian Cochleae. *Journal of the Association for Research in Otolaryngology.* 13:733-745.

[172]. Zachos, F. & Asher, R. (Eds.), 2018. Mammalian Evolution, Diversity and Systematics. (p. 273). In *Handbook of Zoology*. Berlin: de Gruyter.

[173]. Kusuhashi, N., 2013. A new Early Cretaceous eutherian mammal from the Sasayama Group, Hyogo, Japan. *Proc. R. Soc. B* 2013 280, 20130142.

[174]. Zachos, F. & Asher, R. (Eds.), 2018. Mammalian Evolution, Diversity and Systematics. (p. 5, 39). In *Handbook of Zoology*. Berlin: de Gruyter.

[175]. Citelli, R., 1999. Tribosphenic mammal from the North American Early Cretaceous. *Nature* 401:363-366.

[176]. S. Bi, X. Zheng, X. Wang, N. E. Cignetti, S. Yang and J. R. Wible., 2018. An Early Cretaceous eutherian and the placentalBmarsupial dichotomy. *Nature* 558: 390-395

[177]. Zachos, F. & Asher, R. (Eds.), 2018. Mammalian Evolution, Diversity and Systematics. (p. 276). In *Handbook of Zoology*. Berlin: de Gruyter.

[178]. Archibald, J. & Averianov, A., 2015. Mammalian faunal succession in the Cretaceous of the Kyzlkum desert. *Journal of Mammalian Evolution*, 12(1,2): 9-22.

[179]. Archibald, J. & Averianov, A., 2003. The late Cretaceous placental mammal Kulbeckia. *Journal of Vertebrae Paleontology* 23(2): 404-419.

[180]. Averiano, A. And Archibald, J., 2017. Therian postcranial bones from the Upper Cretaceous Bissekty Formation of Uzbekistan. *Proceedings of the Zoological Institute RAS* 321 (4): 443-484.

[181] Kaas, J. 2020. The Organization of the Neocortex in Early Mammals. In Kass, J. ed. *Evolutionary Neuroscience.* (pp. 339-341) London: Academic Press.

[182] Kaas, J. 2020 Evolution of Visual Cortex in Primates. In Kass, J. ed. *Evolutionary Neuroscience.* (pp. 531-348) London: Academic Press.

[183] Kaas, J. et al. 2020. Evolution of Parietal-Frontal Networks in Primates. In Kass, J. ed. *Evolutionary Neuroscience.* (pp.657-667) London: Academic Press.

[184]. Cachel, S., 2015. *Fossil Primates.* (p. 110). Cambridge University Press

[185]. Reis, M. et al., 2018. Using Phylogenomic Data to Explore the Effects of Relaxed Clocks and Calibration Strategies on Divergence Time Estimation: Primates as a Test Case. *Systematic Biology,* 1-23.

[186]. Springer, M. et al., 2012. Macroevolutionary Dynamics and Historical Biogeography of Primate Diversification Inferred from a Species Supermatrix. *PLoS One* 7(11): e49521.

[187]. Chester, S. Et al. 2015. Oldest known euarchontan tarsals and affinities of Paleocene Purgatorius to Primates. *PNAS* 112 (5): 1487-1492.

[188]. Kay, R., 2017. 100 years of primate paleontology. *American Journal of Physical Anthropology* 165:652-676.

[189]. Stephen G,. B. et al. 2017 Oldest skeleton of a plesiadapiform provides additional evidence for an exclusively arboreal radiation of stem primates *Palaeocene R. Soc.* open sci.4170329.

[190]. Cachel, S., 2015. *Fossil Primates.* (p. 123). Cambridge University Press.

[191]. Silcox, M. et al. 2017. The evolutionary radiation of plesiadapiforms. *Evolutionary Anthropology* 26:74-94.

[192]. Bloch, J. et al., 2007. New Paleocene skeletons and the relationship of plesiadapiforms to crown-clade primates. *PNAS* 104(4):1159-1164.

[193]. Silcox, M. et al., 2017. The evolutionary radiation of plesiadapiforms. *Evolutionary Anthropology* 26:74-94.

[194]. Ni, X, Wang, Y.,Hu Y, Li C., 2004. A euprimate skull from the early Eocene of China. *Nature* 427: 65-68.

[195]. Ni, X. et al., 2013. The oldest known primate skeleton and early haplorhine evolution. *Nature* 498: 60-64.

[196]. Beard, K. , 1998. A New Genus of Tarsiidae (Mammalia: Primates) from the Middle Eocene of Shanxi Province, China, with Notes on the Historical Biogeography of Tarsiers. *Bulletin of Carnegie Museum of Natural History* 34: 260-277.

[197]. Beard, K. et al., 1991. First skulls of the early Eocene primate Shoshonius cooperi and the anthropoid-tarsier dichotomy. *Nature* 349:64-67.

[198]. Beard, K. & Wang, J., 2004. The eosimiid primates (Anthropoidea) of the Heti Formation, Yuanqu Basin, Shanxi and Henan Provinces, People's Republic of China. *Journal of Human Evolution* 46: 401-432.

[199]. Gebo, D. Dagosto, M, Beard, K, Qi, T., Wang, J., 2000. The oldest known anthropoid postcranial fossils and the early evolution of higher primates. *Nature* 404: 276-278,

[200]. Zaw, K. et al., 2014. The oldest anthropoid primates in SE Asia: Evidence form LA-ICP-MS U-Pb zircon age in the Late middle Eocene Pondaung Formation, Myanmar. *Gondwana Research* 26: 122-131.

[201]. Gunnel G. & Cichon R., 2008. Revisiting primate postcrania from the Pondaung Formation of Myanmar. In Fleagle JG, Gilbert CC, editors. *Elwyn Simons: A Search for Origins.* (p. 211-228.) New York: Springer.

[202]. Seiffert, E., et al., 2005. Basal Anthropoids from Egypt and the Antiquity of Africa's Higher Primate Radiation. *Science* 310: 300-304.

[203]. Seiffert, E., 2006. Revised age estimates for the later Paleogene mammal faunas of Egypt and Oman. *Proceedings of the National Academy of Sciences* 103(13): 5000-5005.

[204]. Seiffert, E. Simons, E. & Simons, C., 2004. Phylogenetic, biogeographic, and adaptive implications of new fossil evidence bearing on crown anthropoid origins and early stem catarrhine evolution. In C. Ross & R. Kay (eds). *Anthropoid Origins:* New Visions. New York: Kluwer Academic Press.

[205]. Simons, E. et al., 1999. Canine sexual dimorphism in Egyptian Eocene anthropoid primates: Catopithecus and Proteopithecus. *PNAS* 96.5.2559

[206]. Miller, et al., 2017. Patterns of dental emergence in early anthropoid primates from the Fayum Depression, Egypt. *Historical Biology* 30 (1-2): 157-165.

[207]. Simons, E. (2001). The cranium of Parapithecus grangeri, an Egyptian Oligocene anthropoidean primate. *Proceedings of the National Academy of Sciences* 98 (14): 7892-7.

[208]. Bush, E., Simons, E., Almond, J. (2004). High-resolution computed tomography study of the cranium of a fossil anthropoid primate, Parapithecus grangeri: New insights into the evolutionary history of primate sensory systems. *Revue* 281A: 1083-7.

[209] Jacobs, G., 2017. Color Vision in Primates. In (ed.) Fuentes, A. *International Encyclopedia of Primatology.* John Wiley & Sons, Inc.

[210]. McCabe, S. 2017. Aegyptopithecus. in (ed) Fuentes, A. *International Encyclopedia of Primatology.* John Wiley & Sons, Inc.

[211]. Simon, E., Seiffert, E., Rykan, T. & Attia, Y., 2007. A remarkable female cranium of the early Oligocene anthropoid *Aegyptopithecus zeuxis* (Catarrhini, Propliopithecidae). *Proceedings of the National Academy of Sciences* 104 (21): 8731-8736.

[212]. Leutenegger, W. & Cheverud, J., 1985. Sexual Dimorphism in Primates: The effects of Size. In Jungers, W. (ed.) *Size and Scaling in Primate biology*. NY: Springer Science + Business Media.

[213]. Paciulli, L., 2017. Sexual Dimorphism, in (ed) Fuentes, A. *International Encyclopedia of Primatology*. John Wiley & Sons, Inc.

[214]. Miller, E. et al., 2017. Patterns of dental emergence in early anthropoid primates from the Fayum Depression, Egypt. *Historical Biology* Http: dx.doi.orga/0.1080 /089112963.

[215]. Coleman, M., 2007. The functional morphology and evolution for the primate auditory system. Ph.D. State University of New York at Stony Brook, 517:3301482.

[216]. Perry, J. et al., 2017. Articular scaling and body mass estimation in platyrhines and catarrhines: Modern variation and application to fossil anthropoids. *Journal of Human Evolution* https//doi.org/10.1016/J.jhevol.2017.10.008o

[217]. Herculano-Houzel S. et al., 2016. Corticalization of motor control in humans is a consequence of brain scaling in primate evolution. *Journal of Comparative Neurology* 524:448-455.

[218]. Kass, J., 2017. The Evolution of Mammalian Brains from Early Mammals to Present Day Primates. In *Evolution of the Brain, Cognition and Emotion in Vertebrates* Brain Science, DOI 10.1007/ 978-4-431-56559.

[219]. Finarelli J. & Clyde, W., 2004. Reassessing hominoid phylogeny: evaluating congruence in themorphological and temporal data. *Paleobiology* 30(4): 614B651.

[220]. Begun, D., 2017. Evolution of the Hominoidea. in (ed) Fuentes, A. *International Encyclopedia of Primatology*. John Wiley & Sons, Inc.

[221]. Andrews (1981). Species diversity and diet in monkey during the Miocene. *Aspects of Human Evolution*. Taylor & Francis.

[222]. Stevens, N. 2013. Paleontological evidence for an Oligocene divergence between Old World monkeys and apes. *Nature* 497, 611B614,

[223]. Smaers, J. & Soligo, C., 2013. Brain reorganization, not relative brain size, primarily characterizes anthropoid brain evolution. *Proceedings of the Royal Society B* 280: 20130269. http://dx.doi.orgn/ 10.1098/rspb.2013.0269.

[224]. MacLeod, C., Zilles, K., Schleicher, A., Rilling, J., Gibson, K. 2003. Expansion of the neocerebellum in Hominoidea. *Journal of Human Evolution* 24: 401-429.

[225] Smaers, J. et al., 2017. Exceptional evolutionary expansion of prefrontal cortex in Great Apes and Humans. *Current Biology Report* 27: 714-720.

[226]. Barrett, L. Henzi, P., Dunbar, R., 2003. Primate cognition: From "what now?" to "what if?" *Trends in Cognitive Sciences* 7(11): 494-497.

[227] Moorjani, P. et al., 2016. Variation in the molecular clock of primates. *PNAS* 113(38): 10607-10612.

[228] Selby, M & Lovejoy, C., 2017. Evolution of the hominoid scapula and its implications for earliest hominid locomotion. Am J. Phys Anthropol 1-19.

[229] Thompson, N. et al., 2018. Unexpected terrestrial hand posture diversity in wild mountain gorillas. *Am J Phys Anthropol* 1-11.

[230] Cooksey, K. & Morgan, D., 2017. Gorilla (Gorilla). In (ed.) Fuentes, A. *International Encyclopedia of Primatology*. John Wiley & Sons, Inc.

[231] Musgrave, S. & Sanz, C. 2017. Tool Use (Apes). In (ed.) Fuentes, A. *International Encyclopedia of Primatology*. John Wiley & Sons, Inc.

[232] Van Schaik, et al., 2003. Orangutan cultures and the evolution of material culture. *Science* 299: 102-105.

[233]. Finarelli, J. & Clyde, W. 2004. Reassessing hominoid phylogeny: evaluating congruence in the morphological and temporal data (see appendix 3). *Paleobiology*, 30 (4): 614-651.

[234]. Finarelli, J. & Clyde, W., 2004. Reassessing hominoid phylogeny: evaluating congruence in the morphological and temporal data. *Paleobiology* 30 (4): 614-651.

[235]. Begun, D., 2003. Planet of the Apes. *Scientific American*, 74-83.

[236]. Ashely B., 2016. The Phylogenetic position of *Proconsul* and catarrhine ancestral morphotypes. Ph.D. Dissertation, New York University, 284: 10192021.

[237] Nakatsukasa, M., 2019. Miocene Ape Spinal Morphology: The Evolution of Orthogrady. In *Spinal Evolution (Morphology,*

Function, and Pathology of the Spine in Hominoid Evolution). Springer Nature Switzerland.

[238]. Nakatsukasa M. & Kunimatsu, Y., 2009. Nacholapithecus and its importance for understanding hominoid evolution. *Evolutionary Anthropology* 18:103-119.

[239]. Harrison, T., 2017. Micoene Primates. In (Ed.) Fuentes, *The International Encyclopedia of Primatology*. A. John Wiley & Sons.

[240]. Nakatsukasa, M, et al., 2003. *Primates* 44(3): 371-412.

[241]. Andrews, P., 2015. *An Ape's View of Human Evolution*. (p. 56) Cambridge University Press.

[242]. Begun, D., 2016. *The Real Planet of the Apes*. (p. 107) Princeton University Press.

[243]. Cameron, D., 2004. *Hominid: Adaptations and Extinctions*. (p. 92-94). University of New South Wales Press.

[244]. McCrossin, M., 2004. Locomotor diversity among Miocene catarrhines: Another look at retroflection of the medial epicondyle of the humerus. *AAPA* 73rd Annual Meeting, Tampa, FL.

[245]. Nengo, I., 2017. New infant cranium from the African Miocene sheds light on ape evolution. *Nature* 548: 169-174.

[246]. Benefit, B., 2017. Skull secrets of an ancient ape. *Nature* 548:160-161.

[247]. Alba, D., 2017. Ten years in the dump: An updated review of the Miocene primate-bearing localities from Abocador de Can Mata (NE Iberian peninsula) *Journal of Human Evolution*, 102: 12-20.

[248]. Moya-Sola et al., 2009. A unique Middle Miocene European hominoid and the origins of the great ape and human clade. *Proceedings of the National Academy of Sciences* 106(24) 9601-9606.

[249]. Cachel, S., 2015. *Fossil Primates*. (p. 221). Cambridge University Press.

[250] Begun, D., 2017. Evolution of the Hoimoidea. In (ed.) Fuentes, A. *International Encyclopedia of Primatology*. John Wiley & Sons, Inc.

[251]. Moya-Sola, S., Kohler, M. Alba, D, Casanovas-Vilar, I. Galindo, J., (2004). *Pierolapithecus catalaunicus,* a new Middle Miocene Great Ape from Spain. *Science* 306: 1339-1344.

[252] Moya-Sola, S., 2017. Pierolapithecus. In (ed.) Fuentes, A. *International Encyclopedia of Primatology.* John Wiley & Sons, Inc.

[253]. Begun, D., 2005. Comment on "Pierolapithecus catalaunicus, a new Middle Miocene Great Ape from Spain." *Science* 308: 203c

[254] Bohme, M, et al. 2019. A new Miocene ape and locomotion in the ancestor of great apes and humans. *Nature* 575: 489-493.

[255]. Smaers, J. & Soligo, C., 2013. Brain reorganization, not relative brain size, primarily characterizes anthropoid brain evolution. *Proceedings of the Royal Society B* 280: 20130269. http://dx.doi.orgn/10.1098/rspb.2013.0269.

[256]. Almecija, S. & Sherwood, C., 2017. Hands, Brains, and Precision Grips: Origins of Tool Use Behaviors. In *Evolution of Nervous Systems*, 2nd edition, volume 3.

[257]. Fox, M., 2015. The origins of causal cognition in early hominins. *Biol philos,* 30:247-266.

[258]. Kunimatsu, Y. et al., 2007. A new late Miocene ape from Kenya and its implicatins for the origins of African great apes and humans. *PNAS,* 104(49): 19220-19225.

[259]. Bernor, R., 2007. New apes fill the gap. *PNAS,* 104(50) 19661-19662.

[260] Bonis, L. & Koufos, G., 1993. The face and the mandible of Ouranopithecus macedoniensis: description of new specimens and comparisons. *Journal of Human Evolution* 24: 469-491.

[261]. Bonis, L. & Koufos, G., 1994. Our ancestors' ancestor: *Ouranopithecus* is a Greek link in human ancestry. *Evolutionary Anthropology* 3(3): 75-83

[262] Bonis, L. & Koufos, G. 2014. First discovery of postcranial bones of Ouranopithecus macedoniensis (Primates, Hominoidea) from the late Miocene of Macedonia (Greece). Journal of Human Evolution 74:21-36.

[263] Begun, D. 2016 *The Real Planet of the Ape: A New Story of Human Origins*. (p. 179-82). Princeton University Press.

[264]Ioannidou, M., 2019. A new three-dimensional geometric morphometrics analysis of the Ouranopithecus macedoniensis cranium (Late Miocene, Central Macedonia, Greece). Am J Phys Anthropol: 1-13.

[265]. Kohler, M. and Moya-Sola, S., 1997. Ape-like or hominid-like? The positional behavior of *Oreopithecus bambolii* reconsidered. *Proceedings of the National Academy of Sciences* 94: 11747-11750.

[266]. Cameron, D., 2004. *Hominid; Adaptations and Extinctions.* (p. 154.) University of New South Wales Press.

[267] Russo, G. & Shapiro, L., 2013. Revaluation of the lumbosacral region of *Oreopithecus bambolii. Journal of Human Evolution*, 1-13.

[268]. Susman, R., 2004. *Oreopithecus bambolii*: an unlikely case of hominid-like grip capability in a Miocene ape. *Journal of Human Evolution*, 46:105-117.

[269]. Alba, D. et al., 2001. Canine reduction in the Miocene hominoid Oreopithecus bambolii: behavioral and evolutionary implications. *Journal of Human Evolution*, 40:1-16.

[270]. Urciuoli. A. et al., 2020. The evolution of the vestibular apparatus in apes and humans. *Elife* 9:e51261.

[271]. Begun, D., 2016. *The Real Planet of the Apes.* (p. 171). Princeton University Press.

[272] Rook, L., 2017. Oreopithecus. In (ed.) Fuentes, A. *International Encyclopedia of Primatology.* John Wiley & Sons, Inc.

[273]. Lebartard, A. et al., 2008. Cosmogenic nuclide dating of *Sahelanthropus tchadensis* and *Australopithecus bahrelghazali*:

Mio-Pliocene hominids from Chad. *Proceedings of the National Academy of Sciences* 105 (9): 3226-3231.

[274]Ahern, J. 2018 Sahelanthropus. (In ed. Trevathan, W.) *The International Journal of Biological Anthropology.* John Wiley & Sons.

[275]. Brunet, et al., 2002. A new hominid from the Upper Miocene of Chad, Central Africa. *Nature* 418: 145-151.

[276]. Zollikofer, C., Ponce de Leon, M., Lieberman, D., Guy, F., Pilbeam, D., Likius, A., Mackaye, H., Vignaud, P., & Brunet, M., 2005. Virtual cranial reconstruction of *Sahelanthropus tchadensis. Nature* 434: 755-759.

[277].Neauz, D. et al., 2017. Relationship between foramen magnum position and locomotion in extand and extinct hominoids. *Journal of Human Evolution*, 113: 1-9.

[278] **Guy, F. et al. 2021. Postcranial evidence of late Miocene hominin bipedalism in Chad[1]. *Publications du Laboratorie Palevorprim.* (hal-03037386)[2]**

[279]. Novello, A. et al., 2017. Phytoliths indicate significant arboreal cover at Sahelanthropus type locality TM266 in northern Chad and a decrease in later sites. *Journal of Human Evolution*, 106:66-83.

[280] Ahern, J. 2018, *Sahelanthropous, The International Encyclopedia of Biological Anthropology.* Hobokan, NJ: Wiley-Blackwell.

[281]. Benton, M. Et al., 2015. Constraints on the timescale of animal evolutionary history. *Palaeontolgia Electronica*, 18.1.1FC.

[282] Senut, B. 2018. Orrorin. In (Trevathan, W. ed.) *The International Journal of Biological Anthropology.* John Wiley & Sons.

---

1. https://chadcradlehumanity.monsite-orange.fr/file/
   24536468702b0efa22c827f8329496b0.pdf

2. https://hal.archives-ouvertes.fr/hal-03037386

[283]. Senut, B. et al., 2001. First hominid from the Miocene (Lukeino formation, Kenya). C.R. Acad.Sci. Paris, *Sciences de la Terre et des Planetes/Earth and Planetary Sciences* 0: 1-9.

[284]. Galik, K., Senut, B., Pickford, M., Gommery, D., Treil, J., Kuperavage, A., & Eckhardt, R., 2004. External and internal morphology of the BAR 1002'00 *Orrorin tugenensis* femur. *Science,* 305(3): 1450-1453.

[285] Kuperavage, a. et al. 2018. Earliest Known Hominin Calcar Femorale in *Orrion tugenensis* Provides Further Internal Anatomical Evidence for Origin of Human Bipedal Locomotion. *The Anatomical Record* 301 (11): 1834-1839.

[286]. Nakatsukasa, M. Et al., 2007. Femur length, body mass, and stature estimates of Orrorin tugenensis, a 6 Ma hominid from Kenya. *Primates,* 48(3): 171-8.

[287]. Ayala, F. & Cela-Conde, C. (2017). *Processes in Human Evolution.* (p. 130). Oxford University Press.

[288] Pugh, Kelsey D., 2020. The Phylogenetic Relationships of Middle-Late Miocene Apes: Implications for Early Human Evolution. *CUNY Academic Works.*
https://academicworks.cuny.edu/gc_etds/3619.

[289]. Teilhard de Chardin, P., 1961. *The Phenomenon of Man.* NY: Harper and Row.

[290] Susko, M. 2020. *Natural Extravagance and the Dynamic Vulnerability of Life.* Baltimore: AllrOneofUs Publishing.

[291] See *The Generational Patterning of Historical Time*:
https://www.allroneofus.com/the-generational-patterning-of-historical-time.html

Don't miss out!

Visit the website below and you can sign up to receive emails whenever Michael A. Susko publishes a new book. There's no charge and no obligation.

https://books2read.com/r/B-A-GJLJ-BCYEB

BOOKS 2 READ

Connecting independent readers to independent writers.

Did you love *Ten Pulses of Evolution & the Surprising Nature of Evolutionary Time*? Then you should read *Life's Dynamic Vulnerability: A Paradigm Shift in Biology*[3] by Michael A. Susko!

[4]

This work offers new metaphors to understand our biological world, one more in accord with scientific evidence and more in harmony with preserving our planet and leading humans to a fuller life. First, it recognizes a "natural extravagance," that life often exceeds our expectation in terms of number, variety, beauty, and capacity for change. Our ability to understand and even to categorize life's many manifestations is often beyond our reach. Such extravagance is paired with a critical dimension of life, its dynamic vulnerability. Life needs to take in outside energy to exist and is acutely sensitive to the environment with its many vagaries. This vulnerability is paradoxically

---

3. https://books2read.com/u/4NLdgz

4. https://books2read.com/u/4NLdgz

linked to a dynamism which leads to evolutionary novelty. Rather than emphasizing the end state of evolution in terms of "fitness," this work focuses on the vulnerable process of evolution itself. In re-envisioning biology, more accurate to the onrush of discoveries, we offer a vision that better protects and preserves our world. The issue at hand is nothing less than our evolutionary future.

Read more at https://www.allroneofus.com/.

Also by Michael A. Susko

A Couple Through Time
Down Below and the Archon's Castle
Up Above and the Runaway
Across the Gulf and Journey Into Un-Time
On the Bay and a Child Found
In the Wild and Do One Wild Thing
On the Mountain and Two Are Missing
To the Beginning and Journey Through Here

Archetypal Worlds
Alwon in Another World: An Archetypal Voyage
Line On the Wall
The Alien's Gift
The Gold People
Spider Woman and the Timeroc
Quill Ears & the Other Earth
Darkwood and Dual with the Shadow Side
Giant Under the Mountain

Haikus and Photos

Little Lion

Nature Haikus & Photos

The Dreaming Series
Sleek Back
Streak and Cave Bear Dreaming
Moby and Marsupial Mole Dreaming

The Dream World Trilogy
Delphi, the Time Thief, and the Dream World
Detinna and the Cave God
The Resistance & the Empire

Transformational Stories
Caseness and Narrative: Contrasting Approaches to People
Psychiatrically Labelled
Transformative Experiences, Psychiatric Research, and Informed
Consent
Transformational Stories: Voices for True Healing in Mental Health

Writings from Street People
Street Images
Street Images II

Standalone
The Little People & the Time-Rider
Animal Spell

Watch for more at https://www.allroneofus.com/.

About the Author

Holding degrees in Philosophy and Psychology, the author has been dedicated to interdisciplinary, synthetic work for much of his life. Previously, he published in *World Futures*, "The Fragility of Evolution," which reenvisioned evolution around the concept of fragility, and protecting fragile periods of change. Since this work, he offers this important extension to our evolutionary awareness, revealing its surprising logarithmic patterning.

Read more at https://www.allroneofus.com/.

AllrOneofUs Publishing

About the Publisher

AllrOneofUs Publishing seeks out work which will make a novel and qualitative addition to the world literature, and one that will last across generations. Many of these persons are in the later part of their life and have made exemplary contributions which are unrecognized. To cite a few examples, we recommend Rich Mullin's *Ethics and the Full-breasted Richness of Life*, John Susko's *Flowers of the Night: Musings from a Sentimental Son,* and Dr. Curtis Adams' *Psychosis and the Humpty Dumpty Story.*

www.ingramcontent.com/pod-product-compliance
Lightning Source LLC
Chambersburg PA
CBHW021957170526
45157CB00003B/1026